## CABI Regional Tourism Series

by
**Konstantinos Andriotis** Middlesex University, London
and
**Dimitrios Stylidis** Middlesex University, London

Tourism literature includes a plethora of books and conference and journal papers that address various issues of tourism in different regions and subregions of the world. However, up to now there has been a lack of a collection of contribution-based books with a regional and subregional focus. In order to complement the existing tourism literature with new or emerging topics of particular interest, it is the aim of the **CABI Regional Tourism Series** to offer a collection of books with comparative and individual case studies from at least eight regions and subregions of the world. These books will come under the following titles, each of which focuses on some of the countries or country clusters: Tourism in Western Europe; Tourism in Eastern Europe; Tourism in East and Southeast Asia; Tourism in the Middle East; Tourism in Sub-Saharan Africa; Tourism in the Caribbean; Tourism in Latin America; Tourism in Pacific Oceania.

### Titles Available:

**Tourism Planning and Development in Latin America** Carlos Monterrubio, Konstantinos Andriotis and Dimitrios Stylidis
**Tourism Planning and Development in South Asia** Dimitrios Stylidis and Boopen Seetanah
**Tourism Marketing in Western Europe** Nikolaos Boukas and Dimitrios Stylidis
**Tourism Planning and Development in Western Europe** Konstantinos Andriotis, Carla Pinto Cardoso and Dimitrios Stylidis

### Titles Forthcoming:

**Tourism Planning and Development in the Middle East** Stella Kladou, Anna Farmaki, Konstantinos Andriotis and Dimitrios Stylidis
**Tourism Planning and Development in Eastern Europe** Hania Janta, Konstantinos Andriotis and Dimitrios Stylidis

### Suggested Topics:

- Tourism Development and Planning
- Tourism Economics and Finance
- Tourism Marketing, Management and Information Technology

To discuss a proposal for a book in this series, please contact Konstantinos Andriotis at k.andriotis@mdx.ac.uk or Dimitrios Stylidis at d.stylidis@mdx.ac.uk.

### The Series Editors

**Konstantinos Andriotis** is a Professor in Tourism at Middlesex University, London, and in the past he was Associate Professor and Head of the Hotel and Tourism Management Department at the Cyprus University of Technology. He holds an MSc in International Hospitality Management (University of Strathclyde, 1995) and a PhD in Tourism Development and Planning (Bournemouth University, 2000) and was a postdoctoral researcher in Tourism Marketing (Hellenic Open University,

2006). He edits the *International Journal of Tourism Policy*, the *Journal of Qualitative Research in Tourism* and the CABI Regional Tourism Series. He is a member of the editorial board of more than 20 journals. He is author of seven books and over 50 peer-reviewed articles and book chapters. His work has appeared in leading academic journals, including *Annals of Tourism Research*, *Journal of Travel Research*, *Tourism Management* and *Journal of Sustainable Tourism*. He has an h-index of 34 and more than 4700 citations. He was listed in the Stanford University list of the top 2 per cent of scientists in the World. He has acted as chairman and member of the organizing and scientific committees of various international conferences and has given keynote speeches at several conferences. His research interests include tourism development and planning, alternative forms of tourism, tourism experience and degrowth.

**Dimitrios Stylidis** is Associate Professor in Tourism Marketing at the University of Crete, Greece. Prior to this, he was a senior lecturer in Tourism and Research Cluster leader (place image, identity and heritage) at Middlesex University, and a postdoctoral researcher and visiting lecturer at the Ben-Gurion University of the Negev, Israel. He holds a PhD and an MSc in Tourism Management from the University of Surrey, UK. He is the research notes editor of the *International Journal of Tourism Policy*. His work has appeared in leading academic journals including the *Annals of Tourism Research*, *Journal of Travel Research* and *Tourism Management*. His current research interests include place image, tourism impacts and residents' attitudes toward tourism development.

# Tourism Planning and Development in Western Europe

Edited by

**Konstantinos Andriotis**
*Middlesex University, London, UK*

**Carla Pinto Cardoso**
*Universidade Católica Portuguesa, Portugal*

**Dimitrios Stylidis**
*University of Crete, Greece*

**CABI is a trading name of CAB International**

CABI
Nosworthy Way
Wallingford
Oxfordshire OX10 8DE
UK

Tel: +44 (0)1491 832111
E-mail: info@cabi.org
Website: www.cabi.org

CABI
WeWork
One Lincoln St
24th Floor
Boston, MA 02111
USA

T: +1 (617)682-9015
E-mail: cabi-nao@cabi.org



A catalogue record for this book is available from the British Library, London, UK.

Library of Congress Cataloging-in-Publication Data

Names: Andriotis, Konstantinos, editor. | Cardoso, Carla, editor. | Stylidis, Dimitrios, editor.
Title: Tourism planning and development in Western Europe / Edited by: Konstantinos Andriotis, Middlesex University London, UK, Carla Cardoso, Universidade Católica Portuguesa, Portugal, Dimitrios Stylidis, University of Crete, Greece.
Description: Boston : CAB International, [2022] | Series: Cabi regional tourism series | Includes bibliographical references and index. | Summary: "Focusing exclusively on tourism development, planning and policy, this book draws together new voices to provide case studies illuminating issues across western Europe. It adopts a range of methodological approaches to cover areas such as industrial tourism, natural and cultural heritage, and micro-destinations, and considers post-COVID tourism"-- Provided by publisher.
Identifiers: LCCN 2021048932 (print) | LCCN 2021048933 (ebook) | ISBN 9781800620797 (hardback) | ISBN 9781800620803 (ebook) | ISBN 9781800620810 (epub)
Subjects: LCSH: Tourism--Europe, Western--Management. | Tourism--Europe, Western--Planning.
Classification: LCC G155.E85 T686 2022 (print) | LCC G155.E85 (ebook) | DDC 338.4/7914--dc23/eng/20211216
LC record available at https://lccn.loc.gov/2021048932
LC ebook record available at https://lccn.loc.gov/2021048933

References to Internet websites (URLs) were accurate at the time of writing.

ISBN-13: 9781800620797 (hardback)
9781800620803 (ePDF)
9781800620810 (ePub)

DOI: 10.1079/9781800620797.0000

Commissioning Editor: Claire Parfitt
Editorial Assistant: Lauren Davies
Production Editor: Tim Kapp

Typeset by Exeter Premedia Services Pvt Ltd, Chennai, India
Printed and bound in the USA by Integrated Books International, Dulles, Virginia

# Contents

# Contributors

**Helena Albuquerque** is an Assistant Professor in the Department of Tourism, Heritage and Culture, and member of REMIT research unit (Research on Economics, Management, and Information Technologies) of Portucalense University, Portugal, being responsible for the line Tourism, Hospitality and Heritage. She also collaborates with GOVCOPP (Research Unit on Governance, Competitiveness and Public Policies) of Aveiro University, Portugal. She has participated in several research projects in the areas of sustainable tourism and environmental conservation. Her main research interests are Sustainable Tourism, Geographic Information Systems applied to Tourism and Nature Tourism. E-mail: helenaa@mail.upt.pt

**Sonia Ancio Alcón** is a degree student in Tourism at the University of Seville, Spain. She has worked as an internal student of the Department of Applied Economics at the University of Seville, showing interest in issues related to tourism, especially the development of industrial tourism and the Andalusian economy. E-mail: anadq@us.es

**Konstantinos Andriotis** is a Professor in Tourism at Middlesex University, London. He holds a PhD in Tourism Development and Planning (Bournemouth University, 2000) and was a postdoctoral researcher in Tourism Marketing (Hellenic Open University, 2006). He edits the 'International Journal of Tourism Policy', the 'Journal of Qualitative Research in Tourism' and the CABI Regional Tourism Series. He has an h-index of 34 and more than 4700 citations. He was listed in the Stanford University list of top 2 per cent of scientists in the world. His research interests include tourism development and planning, alternative forms of tourism, tourism experience and degrowth. E-mail: k.andriotis@mdx.ac.uk

**Ágúst Bogason** is an environmental social scientist with an academic background in political science and social anthropology. He works with policy and governance, sustainable rural development, the labour market, local participation in policy processes and tourism studies, mainly in the rural Nordic countries and Arctic regions. His research often focuses on the intersection between research, policy, practice and implementation. Ágúst works as a research fellow at Nordregio, a research centre on regional development, policy and planning established by the Nordic Council of Ministers. E-mail: agust.bogason@nordregio.org

**Rikke Brandt Broegaard** is a human geographer with a PhD in international development studies. She works in the fields of community development, rural areas, natural resource management and inequality, in Denmark, Scandinavia and in the Global South. In a Nordic context, her research focuses on rural development and peripheralization processes, including place-making and translocal connections, with special attention to the role of micro-entrepreneurs, including in the tourism sector. She works at the Centre for Regional and Tourism Research, Bornholm, and at the Institute for Geoscience and Natural Resource Management, Copenhagen University. E-mail: rikke.brandt.broegaard@crt.dk

**Gildas Buron** is Head of the Museum and Heritage Department of CAP Atlantique and Curator of the Salt Marshes Museum (Batz-sur-Mer, France). A graduate in history (Poitiers), he has been conducting research on traditional salt works, production and trade in Brittany for 35 years. His main interests are economic and social history and the history of landscapes. He has published two illustrated books and several dozen articles. He has taken part in several national and international projects on the history of salt and salt marshes, where he shares his field observations and his experience as a heritage and tourism player. E-mail: gildas.buron@cap-atlantique.fr

**Carla Pinto Cardoso** holds a PhD in Economics of Tourism at Bournemouth University (UK) and is the coordinator of the Tourism and Heritage Department in the Faculty of Philosophy and Social Sciences at Universidade Católica Portuguesa (Portugal), where she also teaches in the tourism and economic fields. She is member of the R&D Centre for Philosophical and Humanistic Studies and her research and publication interests include tourism impacts, planning and strategic management. While working as senior lecturer and department coordinator she supervised a number of undergraduate projects, masters dissertations and theses. She has published in a number of journals and participated in and organized several conferences. In 2019, she was chair of the International Conference on Tourism – ICOT 2019. She works as a senior tourism consultant, involved in national and international tourism projects. E-mail: cpintocardoso@ucp.pt

**Margarida Ferreira da Silva** is a Foundation for Science and Technology (FCT) research fellow, with a PhD thesis devoted to tourism and coastal and maritime cultural heritage. She is also a team member of H2020 PERICLES – Preserving and Sustainably Governing Cultural Heritage and Landscapes in European Coastal and Maritime Regions project and a member of GOVCOPP (Research Unit on Governance, Competitiveness and Public Policies) and CESAM (Centre for Environmental and Marine Studies) (University of Aveiro). Over recent years she has been involved in several national and international projects related to sustainable tourism and coastal and maritime cultural heritage. E-mail: a.silva@ua.pt

**Valeria Diaferia** holds a graduate degree in Foreign Languages and Literatures at University Aldo Moro in Bari and continues her studies with a master's in International Tourism and Event management at Hochschule Fresenius University of Applied Sciences in Cologne, Germany. Thanks to some international experience and in particular an internship in Dublin, she has been increasingly fascinated by the wide world of tourism, in particular destination resilience and its sociocultural dimension, together with its implications for innovative product design in experimental destinations in Germany, Italy and Ireland. E-mail: valeriadiaferia7@gmail.com

**Ana M. Domínguez-Quintero** has a PhD in Economics and is an Associate Professor of Applied Economics at the University of Seville, Spain. Ana's research interests focus mainly on cultural tourism. Her dissertation, 'The development of a structural model to measure satisfaction in cultural tourism' obtained a European Mention accreditation at the University of Seville. She has published papers on topics related to authenticity, quality of experience, emotions and satisfaction. She has been a guest professor at York St John University (UK). Her academic activities at the University of Seville are related to teaching quantitative and qualitative methods. E-mail: anadq@us.es

**Martin Fontanari** is a professor in Tourism and Event Management at the ISM – International School of Management – in Cologne. He holds a PhD in Business Administration (University of Trier, 1994), after Master studies in Innsbruck (1989) and research studies at the ISBM at PennState University and Wharton Business School (1990). Since joining the tourism consultancy business more than 20 years ago, he has managed 150 projects worldwide in the field of destination marketing and destination development and contributed more than 70 publications. Currently his focus is on developing resilient destinations, overcoming overtourism and post-covid-19 strategy approaches. E-mail: martin.fontanari@ism.de

**Valeria Fusè** is an assistant producer at FreedaMedia, a digital and social media start-up based in Milan. Fusè graduated in 'Science of Tourism and Local Community from the University of Milan-Bicocca (ITA), and holds a master's in Planning and Management of Tourism Systems from the University of Bergamo (ITA). Fusè's master's thesis, 'Host–guest relationships in urban tourism: a comparative analysis between Milan and Porto' was partly conducted in Portugal during a research stay at the University of Aveiro (PRT), financed by a scholarship. Fusè participated in the Milan-Bicocca Summer School at the BRIC at University of Chongqing (CHN), the 'Erasmus+' programme at Tampere University of Applied Sciences (FIN) and the 'Real Cities | Responsive Environment' research project in collaboration with the Harvard Graduate School of Design (MA-USA), financed by a GITT scholarship. E-mail: valeriafuse099@gmail.com

**Anna Karlsdóttir** is a human geographer and holds a PhD in Social Sciences from Roskilde University, Denmark. She is currently a senior research fellow at Nordregio and has held an associated professorship in Human Geography and Tourism Studies at the University of Iceland since 2002 (with leave of absence until 2023). Her research focuses on transformation of the labour market, occupational shifts and transformations in the Nordic and Arctic region, employment diversification, tourism and related creative industries and occupations linked to regional development. Her focus has also been gender, youth, labour migration and cruise tourism, as well as monitoring tourism in the High North. E-mail: anna.karlsdottir@nordregio.org

**Elisabeth Kastenholz** is an Associate Professor at the University of Aveiro, Portugal, where she teaches tourism- and marketing-related subjects. She coordinates the university's research unit GOVCOPP and serves as director at the international Doctoral Program in Marketing and Strategy. She holds a PhD in Tourism Studies, an MBA, a graduate course in Tourism Management and Planning and a bachelor's degree in Public Administration/Foreign Affairs. She has researched and published extensively and supervised numerous masters and PhD theses in tourist consumer behaviour and destination marketing, with particular interest in sustainable tourism development, the tourist experience, accessible tourism, rural, nature-based and wine tourism. E-mail: elisabethk@ua.pt

**Marie-Louise Mangion** is a senior lecturer in the Department of Public Policy at the Faculty of Economics, Management and Accountancy at the University of Malta. She holds a PhD from the University of Nottingham. For over 20 years she worked in the public sector, including for the Malta Tourism Authority. She directed tourism research, sourced and managed European funding for the sector, worked on European affairs and formulated tourism and environment policies. On behalf of UNEP, she delivered training on carrying capacity assessment of tourism in various countries. Her research interests include policy design, evidence-based policymaking, tourism policy, tourism economics, development and competitiveness. E-mail: marie-louise.mangion@um.edu.mt

**Filomena Martins** is an Associate Professor at the University of Aveiro, in the Department of Environment and Planning. She is vice-director of the Marine and Atmospheric Sciences Master (second cycle, Bologna), director of the Doctoral Program in Environmental Sciences and Engineering and coordinator of the Sea Branch of the UACoopera. She participated in several European and national research projects in the fields of coastal management and planning, sustainable development, conservation of natural and cultural heritage, risk perception and resilience (combining natural, environmental, spatial and social dimensions).

Her main research interests include coastal and marine planning, governance and nature heritage knowledge. E-mail: filomena@ua.pt

**Laura Rauschen** was born in 1997 in East Belgium. In 2017, she received her bachelor's degree in Hotel Management in Liège, Belgium. In 2020, she received her master's degree in Tourism and Event Management from Hochschule Fresenius University of Applied Sciences in Cologne, Germany. Her contribution to Chapter 6 in this book is based on her master's thesis 'Strategic considerations for sustainable tourism development of the micro-destination East-Belgium'. She is interested in sustainable tourism and is curious about how her home region of East Belgium will develop sustainably in the following years. E-mail: laura.rauschen@gmail.com

**Dimitrios Stylidis**, PhD, is Associate Professor in Tourism Marketing at the University of Crete, Greece. Prior to this, he was a senior lecturer in Tourism and Research Cluster leader (place image, identity and heritage) at Middlesex University, and a postdoctoral researcher and visiting lecturer at the Ben-Gurion University of the Negev, Israel. Dimitrios holds an MSc and PhD in Tourism Management from the University of Surrey, UK. His current research interests include place image and marketing, tourist destination image and behaviour, tourism impacts and residents' attitudes towards tourism development. His work has been published in several journals including *Tourism Management*, *Annals of Tourism Research*, *Journal of Travel Research*, *Journal of Sustainable Tourism* and *Journal of Travel and Tourism Marketing*. E-mail: d.stylidis@uoc.gr

**Anastasia Traskevich** studied International Tourism and Economics. She worked as a project manager, a management consultant and tourism expert for spa and wellness enterprises and tour operators in Belarus and Russia. In 2016, she completed her doctoral degree on the topic of competitiveness assessment and development for the health and wellness tourism industry with the highest distinction. She then became a tourism lecturer in the Belarusian State Economic University, Minsk and Hochschule Fresenius, Cologne. The areas of her academic interest are the fields of tourism management, resilience in tourism, the spa industry, and health and wellness tourism. E-mail: anastasia.traskevich@akalem.de

**Anna Serena Vergori**, PhD, is Assistant Professor in Economics at the University of Salento, Italy. Her main research interests concern the impact of tourism on regional development, the relationship between tourism and transportation, seasonality of tourism demand and its determinants in the short and medium-term. Another field of interest is the measurement of innovation in the service sector. Time series analysis is the main econometric method used in her research. E-mail: serena.vergori@unisalento.it

**Peter Wiltshier**, PhD, was a senior lecturer at the University of Derby for 16 years. He is now a research consultant in New Zealand. His interest in community regeneration comes from an abiding concern for the role of the resident and host in local communities and ways in which visitors are accorded a welcome, ensuring that the public and private sectors work together to develop resources and skills. It is the pursuit of bottom-up planning and policy development that is sought and is to be enabled through his work in consultancy and teaching. E-mail: pwiltshier@yahoo.co.nz

# Introduction

**Konstantinos Andriotis[1] and Carla Pinto Cardoso[2]**
*[1]Middlesex University, London, UK; [2]Universidade Católica Portuguesa, Portugal*

While the Covid-19 (coronavirus) pandemic has resulted in a tremendous decline in tourism numbers, with international arrivals down 83% in the first quarter of 2021 compared to the same period in 2020 (UNWTO, 2021a), tourism continues to be one of the largest economic sectors worldwide. Indeed, even the figures nowadays are still far from the results observed in 2019, when tourism contributed 10.3% to global GDP (5.5% in 2020) and employed 330 million people across the world (272 million in 2020). Tourism's contribution over time has made it a fundamental pillar of the world economy, to which no country or region can remain indifferent.

In the global tourism arena, Europe is the most visited region, welcoming half of the world's international tourist arrivals every year. International tourism receipts represent approximately 37% of worldwide tourism earnings (UNWTO, 2020). Moreover, five out of the top ten visited destinations in the world are located in Europe, and include France, Spain, Italy, Germany and the United Kingdom.

Europe can be divided into several geographic regions. However, a broad grouping that has been widely used for historical and political purposes divides Europe into two broad regions: Western Europe and Eastern Europe (University of Minnesota, 2012). While this traditional grouping might not be as relevant today as it has been historically, it is used for the purpose of the CABI Regional Tourism series and includes in total 23 countries: Austria, Belgium, Denmark, Finland, France, Germany, Gibraltar, Greece, Iceland, Italy, Ireland, Liechtenstein, Luxembourg, Malta, Monaco, Norway, the Netherlands, Portugal, San Marino, Spain, Sweden, Switzerland and the United Kingdom. Following the withdrawal of the United Kingdom from the European Union on 31 January 2020, most of the Western European countries (15 out of 23) belong to the European Union which currently counts 27 member states sharing a single market.

Table 1 illustrates the key tourism statistics for each Western European country where data are available.

These statistics highlight the economic importance of tourism for Western European countries, prior to the pandemic. In fact, it becomes evident that the travel and tourism sector represents a major part of gross domestic product (GDP) for many Western European economies. Also, these figures reflect that this sector is a significant source of foreign exchange and tax revenue.

On average for Western countries (excluding Gibraltar, Liechtenstein, Monaco and San Marino, due the lack of available data), the total contribution of the travel and tourism industry, in 2019, represented 10.3% of GDP. This proportion is largest in the Mediterranean countries, such as Greece, where tourism accounts for 20.3% of GDP, followed by Portugal (17.1%), Spain (14.1%), Malta (15.9%) and

 *Tourism Planning and Development in Western Europe*
(eds K. Andriotis *et al.*)
DOI: 10.1079/9781800620797.intr

**Table 1.** Key economic tourism and travel statistics for Western European countries (2019). (WTTC, 2019, 2020 [by country]; UNWTO, 2020, 2021b).

| | Total contribution of travel & tourism (T&T) to GDP* | | Total contribution of T&T to employment* | | International visitor impact | |
|---|---|---|---|---|---|---|
| Country | Receipts Total (USD) | % | Jobs (000s) | % of total employment | Receipts USD (million) | Visitors spend of total exports (%) |
| Austria | 50.4$_{BN}$ | 11.0 | 529.7 | 11.7 | 26.0$_{BN}$ | 10.2 |
| Belgium | 30.5$_{BN}$ | 5.6 | 318.3 | 6.5 | 10.7$_{BN}$ | 2.4 |
| Denmark | 23.1$_{BN}$ | 6.4 | 190.4 | 6.8 | 9.7$_{BN}$ | 4.6 |
| Finland | 21.7$_{BN}$ | 7.8 | 221.9 | 8.3 | 5.6$_{BN}$ | 5.0 |
| France | 240.5$_{BN}$ | 8.5 | 2,691.6 | 9.5 | 69.0$_{BN}$ | 7.7 |
| Germany | 393.1$_{BN}$ | 9.8 | 5,872.6 | 13.0 | 53.4$_{BN}$ | 2.8 |
| Gibraltar | n/a | n/a | n/a | n/a | n/a | n/a |
| Greece | 43.5$_{BN}$ | 20.3 | 824.7 | 21.1 | 23.2$_{BN}$ | 26.9 |
| Iceland | n/a | n/a | n/a | n/a | n/a | 11.9 |
| Italy | 269.8$_{BN}$ | 13.1 | 3,499.3 | 15.0 | 52.1$_{BN}$ | 8.0 |
| Ireland | 17.5$_{BN}$ | 4.2 | 139.0 | 6.0 | 14.3$_{BN}$ | 2.7 |
| Liechtenstein | n/a | n/a | n/a | n/a | n/a | n/a |
| Luxembourg | n/a | 9.1 | n/a | n/a | 6.2$_{BN}$ | 4.7 |
| Malta | 2,451.9$_{MN}$ | 15.9 | 54.4 | 21.3 | 2,198.8$_{MN}$ | 9.9 |
| Monaco | n/a | n/a | n/a | n/a | n/a | n/a |
| Netherlands | 101.6$_{BN}$ | 10.8 | 1,415.6 | 18.1 | 25.0$_{BN}$ | 3.2 |
| Norway | 28.7$_{BN}$ | 7.9 | 310.8 | 11.4 | 6.2$_{BN}$ | 4.7 |
| Portugal | 42.8$_{BN}$ | 17.1 | 1,003.7 | 20.7 | 25.7$_{BN}$ | 23.6 |
| San Marino | n/a | n/a | n/a | n/a | n/a | n/a |
| Spain | 202.1$_{BN}$ | 14.1 | 2,855.7 | 14.4 | 89.8$_{BN}$ | 17.8 |
| Sweden | 40.2$_{BN}$ | 7.3 | 384.7 | 7.5 | 11.3$_{BN}$ | 4.3 |
| Switzerland | 57.0$_{BN}$ | 7.4 | 493.5 | 9.7 | 22.1$_{BN}$ | 5.1 |
| United Kingdom | 305.0$_{BN}$ | 10.1 | 4,268.2 | 12.0 | 45.7$_{BN}$ | 4.9 |
| Average | – | 10.3 | – | 12.2 | – | 8.4 |

n/a = figures not available for 2019 and 2018
*These figures concern direct, indirect and induced impacts.

Italy (13.1%). However, figures also illustrate that there are significant differences between countries. For countries such as Ireland, Belgium and Denmark the total contribution of travel and tourism to GDP is just under 6.5%, despite the importance that the sector has to the general health of these economies. In France, the world's top tourism destination, travel and tourism represents 8.5% of GDP. On the other hand, a closer look at these figures by total receipts reveals that the largest economies in terms of travel and tourism contribution to GDP, in 2019, were Germany (US$393.1 billion), the United Kingdom (US$305.0 billion), Italy (US$269.8 billion), France (US$240.5 billion) and Spain (US$202.1 billion).

In terms of jobs and labour volume, the travel and tourism industry accounts for more than 25 million European jobs (including, indirectly, jobs supported by the industry).

Germany, the United Kingdom, Italy, Spain and France account for near three quarters of the total tourism employment in the Western Europe. Comparing both travel and tourism contributions to the European labour force and to Western European GDP, figures from Table 1 show that the contribution to the labour force is slightly higher than the share of GDP and stands at an average of 12.2% (against 10.3%). Additionally, most countries with an above average share of travel and tourism in GDP also exhibit an above average share of employment. For instance, the Mediterranean destinations of Malta, Greece and Portugal have by far the highest share of their labour force in the travel and tourism sector (21.3%, 21.1% and 20.7%, respectively), well illustrating the significance of the sector to the income of the citizens of these three economies, who highly depend on it.

In terms of visitor impacts, as can be observed from Table 1, in 2019, Western European countries earned more than US$500 billion from international tourism receipts, with Spain (89.8 billion), France (69.0 billion), Germany (53.4 billion), Italy (52.1 billion) and the United Kingdom (45.7 billion) ranked among the top earners. Thus, it is not surprising that Western Europe has the biggest share of global international tourism spending.

When analysing visitor spending relative to total exports, the data show that, on average, international tourism receipts account for 8.4% of overall exports of goods and services. However, it is also evident that there is substantial variation across the countries, ranging from 2.4% in Belgium to 26.9% in Greece. This is a clear indication that some economies are more dependent on visitors' consumption than others.

To sum up, the above figures reveal that tourism has an enormous capacity to provide wealth and employment in Western countries. It is also evident that tourism is undeniably an important driver of economic and social development in Western Europe, notwithstanding the fact that it generates rather mixed results across Western destinations.

In this context, it is not surprising that, while shared socioeconomic, cultural and political conditions of subregions within the Western European region have led some countries in the region to share tourism development experiences, most Western European countries have undertaken diverse pathways in tourism development and planning, with most of them having experienced fast development and growth over the last decade. However, it should be noted that at some point in their development process, some of these destinations began to experience overtourism. Overtourism occurs where there are too many visitors to a particular destination. For instance, due to unlimited growth and associated overtourism, residents of several European cities, such as Venice, Barcelona, Amsterdam and Santorini, were suffering (Coldwell, 2017; Boluk *et al.*, 2019; Milano *et al.*, 2019; Valdivielso and Moranta, 2019; Fontanari *et al.*, 2021). As a result, various policies are required by communities and policymakers to make responsible decisions that respect the limits to growth of each respective community and result in the sustainability of their resources (Andriotis, 2002, 2018).

During the Covid-19 pandemic, the well-established Western European tourism industry that for several years had been grappling with concerns of overtourism, within a short time moved from overtourism to nontourism and it is expected that the economic loss will last for many years into the future. Covid-19 has created a significant impact on international travel and tourism in Western Europe (Aerospace Technology, 2020). Indeed, in line with the world trend, the pandemic turned the tourism industry upside down in Western Europe from early 2020, putting an end to ten years of dynamic growth. Western Europe has been one of the hardest hit regions in the world. The Covid-19 outbreak was an unexpected shock for the global tourism industry, especially as the fear factor restricted tourists' mobility and governments closed their borders in their efforts to curtail and control the pandemic (Andriotis, 2021).

According to data observed by the World Tourism Barometer, published in July 2021, over the first five months of 2021, the number of international tourism arrivals in Europe was 85% below 2019 levels, making it the second largest region in the world affected by Covid, only preceded by Asia (UNWTO, 2021a). This collapse in tourism demand threatened millions of jobs in the sector, especially in Mediterranean countries where the proportion of jobs generated by tourism is higher (e.g. Spain, Greece,

Malta and Portugal), as well as in those countries where the number of international visitors is higher, such as France, the United Kingdom, Spain and Italy. Since May 2021, international tourism has experienced slightly better results, as some governments have started to ease restrictions for vaccinated travellers and as consumer confidence has risen slightly (UNWTO, 2021a). To what extent international visitors will return to their pre-pandemic 2019 levels remains a major uncertainly. However, most UNWTO experts foresee a return to those levels by 2023 (at best) (UNWTO, 2021b). This is essentially attributable to vaccination programmes around the world and the lifting of travel restrictions.

Within this context and despite the significance of tourism for the region of Western Europe, there are limited book collections. The most recent was edited by Pellejero Martínez and Luque Aranda (2020). This book explored the tourism phenomenon in the region and included contributions from 11 expert researchers. The book dealt with the history and growth of tourism in contemporary post-war Western Europe, especially during the periods following the First and Second World Wars and the Spanish Civil War.

A second book edited by Voase (2002) provided an interesting collection of stories authored by a variety of writers with strong local ties to the place of their interest. The book communicated changes in tourism development and practices and reflected how tourism development sought new ways of tourism thinking. The book identified demographic, environmental and consumer trends that Voase believed would influence tourism in Western Europe in the face of 21st-century issues such as ageing, global warming and active and passive consumer segments, and considered how these will influence the future of tourism in different regions.

Williams and Shaw (1999), in *Tourism and Economic Development: Western European Experiences*, highlighted the need for comparative international tourism research and presented a collection of chapters from ten countries. These chapters were written by specialists in tourism and cover changes in the domestic and foreign tourism industry in Western Europe in relation to seasonal and regional concentrations as well as the economic implications of tourism development and policies.

Given the huge significance of tourism as an economic, social, cultural and environmental phenomenon in Western Europe, and the fact that published books focusing on tourism in Western Europe either offer a historical perspective or are outdated, this collection focuses exclusively on tourism development, planning and policy in a wide range of Western European countries and considers various scenarios of post-Covid tourism in Western Europe. The book aims to provide a thorough analysis of the nature of tourism development and planning in Western European countries as well as to document and explore various contemporary issues including social, environmental and cultural aspects, and their implications for sustainable tourism development. Thus, the book will advance understanding of the positive and negative impacts of tourism development and its implications for local communities.

This edited volume includes an introduction, six chapters dealing with tourism development in one particular country (or in the case of chapter 5, two), three chapters analysing tourism between countries or cities in the region and a conclusion. In total, the book documents experiences, challenges, successful and unsuccessful stories, specific cases and other tourism-related issues in Western Europe. Each chapter is authored by scholars who have done extensive research on tourism in their respective countries.

In Chapter 1, Peter Wiltshier offers a critical analysis of practices adopted for the implementation of sustainable development goals in the English Midlands. Through the lenses of various discourses, this case study epitomizes a long-held belief in the positive outcomes of rejuvenation and regeneration following the decline in manufacturing and extractive industry that occurred in the 1980s. The author performs several evaluations, such as relational discourse, structuration, endogeny, stakeholders and structure agency/actor networks, and provides recommendations for any destination, especially those more mature places that need skills, capacity and resources to effect change for future welfare and an enduring legacy. He concludes that for the benefit of the majority of stakeholders, institutional innovation and a partnership approach are required through a process of collaboration.

Chapter 2, by Anna Serena Vergori, explores international tourism in Italy in the last 20 years to investigate the relationship between transportation and tourism. By using a database from the Bank of Italy (years 1997 to 2019), the author analyses foreign tourists' transport modal choices to reach Italian destinations and evaluates their impact on tourism demand seasonality. The main findings include that air transport is preferred, mainly due to low-cost carrier options and that tourists using cars tend to travel more during the peak season compared to those travelling by plane. The study concludes that policymakers, when aiming to reduce seasonality, should consider the relationship between transport modes and seasonality and adopt appropriate strategies.

In Chapter 3, Marie Louise Mangion reports on the tourism experience of Malta. The chapter reveals that although Malta is geographically one of the smallest destination countries within Western Europe, it has a particular advantage in relation to the development of its tourism industry. Through content analysis of several published and unpublished policy documents and scholarly studies, the author identifies key impacts, challenges and contrasts that Malta faces in the development of its tourism industry. Finally, the chapter presents a conceptual framework that depicts several factors supporting or undermining short- and long-term tourism development in a European island context. These factors are characterized by economic dependence on tourism, high levels of population density and urbanization.

In Chapter 4, Ana M. Domínguez-Quintero and Sonia Ancio Alcón analyse industrial tourism in the province of Seville, Spain, which according to the authors emerged at the beginning of the 21st century as a form of cultural tourism and as a response to globalization processes. The authors conclude that firms offering industrial tourism activities obtain important benefits, including direct promotion, increased sales, greater customer satisfaction and an improvement of the image and positioning of the company. All these benefit the local economy of the areas under study.

Chapter 5 by Margarida Ferreira da Silva, Helena Albuquerque, Filomena Martins, and Gildas Buron conducts an explorative comparative study between Aveiro, Portugal and Guérande, France, to explore the unique cultural and natural landscape that salt pans offer to visitors. Due to this landscape, visitors to salt pans combine tourism activities with artisanal salt production. The authors critically analyse the duality between traditional activity disappearance risk and new market trends and propose to policymakers and other stakeholders sustainable integrated strategic guidelines on the potential of salt pans to create multifunctional synergies.

Laura Rauschen, Anastasia Traskevich and Martin Fontanari, in Chapter 6, use East Belgium as a case to conceptualize the relevance of micro-destinations for tourism development and strategic planning. The authors provide an integrative definition of the term 'micro-destination' and explore the extent to which the region of East Belgium can be called a micro-destination. In addition, they pay attention to the strategic considerations required to ensure the sustainable tourism development of the region. By collecting primary data through a workshop and expert interviews the authors propose projects that will only be efficiently implemented if a partnership and team spirit of all stakeholders is ensured.

Chapter 7, authored by Valeria Fusè and Elisabeth Kastenholz, presents a comparative study between Milan (Italy) and Porto (Portugal) to analyse the perceived tourism impacts of these two urban destinations on host communities. The authors explore resident–tourist interactions from the residents' perspective and, by using a mixed-methods approach, attempt to show that residents tend to have different modes of interaction with and hold diverse attitudes to tourism development. In particular, in Porto tourism is perceived as a main driver of change in the old town that generates both positive and negative attitudes, while residents of Milan perceive that while the city is less dependent on tourism development, tourism is integrated in the general transformation of the city and takes place in a more dispersed, spatial manner. As a result, the authors conclude that host–guest encounters occur more frequently in the historical centre of Porto compared to Milan.

In Chapter 8, Valeria Diaferia, Anastasia Traskevich and Martin Fontanari use as cases the countries of Germany, Italy and Ireland to explore the relevance of the sociocultural dimension of destination resilience and its implication for innovative product design. By doing so, they aim to offer new insights for the sustainable and responsible development of tourism in the three countries under study. They use qualitative and quantitative indicators (emotions and creativity; learning skills and professional efficiency; inner and outer world balance; communication and positive thinking; ethics and values; spirituality) to propose implications focusing on enhancing experimental destinations and positive response to change.

In the case of rural areas in Nordic countries (Denmark, Finland, Iceland, Norway, Sweden and the Faroe Islands, Greenland and Åland), Rikke Brandt Broegaard, Ágúst Bogason and Anna Karlsdóttir, in Chapter 9, undertake a content analysis of 110 rural, subnational (i.e. municipal, subregional and regional) tourism plans to explore several visions and goals for tourism development. The authors use predefined and emergent coding categories, condensations and matrix queries, to explore the extent to which regional tourism strategies consider sustainability concerns. Finally, they provide an understanding of the degree to which tourism actors, policymakers and local communities in Nordic countries employ various regional tourism strategies in order to balance economic development with environmental and social pressures.

The concluding chapter, written by Dimitrios Stylidis, reviews the previous chapters and outlines the main reasons that have made planning and policy essential to tourism development and planning in Western Europe. The main conclusion drawn is that sustained growth will be instrumental in supporting the economic recovery of many Western European countries following the Covid-19 pandemic. The chapter also provides directions for future research.

## References


Aerospace Technology (2020) Tourism in Western Europe has been hit hard by COVID-19. Available at: https://www.aerospace-technology.com/comment/tourism-in-western-europe-has-been-hit-hard-by-COVID-19/ (accessed 12 June 2021).

Andriotis, K. (2002) Options in tourism development: Conscious versus conventional tourism. *Anatolia* 13(1), 73–85. DOI: 10.1080/13032917.2002.9687016.

Andriotis, K. (2018) *Degrowth in tourism: Conceptual, theoretical and philosophical issues*. CAB International, Wallingford, UK.

Andriotis, K. (2021) Introduction. In: Andriotis, K. (ed.) *Issues and Cases of Degrowth in Tourism*. CAB International, Wallingford, UK, pp. 1–21.

Boluk, K.A., Cavaliere, C.T. and Higgins-Desbiolles, F. (2019) A critical framework for interrogating the United Nations Sustainable Development Goals 2030 Agenda in tourism. *Journal of Sustainable Tourism* 27(7), 847–864. DOI: 10.1080/09669582.2019.1619748.

Coldwell, W. (2017) First Venice and Barcelona: now anti-tourism marches spread across Europe. *The Guardian,* 10 August 2017.

Fontanari, M., Traskevich, A. and Seraphin, H. (2021) (De)growth imperative: The importance of destination resilience in the context of overtourism. In: Andriotis, K. (ed.) *Issues and Cases of Degrowth in Tourism*. CAB International, Wallingford, UK, pp. 22–41.

Milano, C., Novelli, M. and Cheer, J.M. (2019) Overtourism and degrowth: a social movements perspective. *Journal of Sustainable Tourism* 27(12), 1857–1875. DOI: 10.1080/09669582.2019.1650054.

Pellejero Martínez, C. and Luque Aranda, M. (eds) (2020) *Inter and Post-war Tourism in Western Europe, 1916–1960*. Palgrave Macmillan, Cham, Switzerland.

University of Minnesota (2012) World Regional Geography. Available at: https://open.lib.umn.edu/worldgeography/chapter/2-3-regions-of-western-europe/ (accessed 15 June 2021).

UNWTO (2020) *UNWTO World Tourism Barometer*, 18(1), January 2020.

UNWTO (2021a) Tourist Numbers Down 83% but Confidence Slowly Rising. Available at: https://www.unwto.org/taxonomy/term/347 (accessed 12 July 2021).

UNWTO (2021b) World Tourism Barometer, July 2021. Available at: https://www.unwto.org/unwto-world-tourism-barometer-data (accessed 22 July 2021).

Valdivielso, J. and Moranta, J. (2019) The social construction of the tourism degrowth discourse in the Balearic Islands. *Journal of Sustainable Tourism* 27(12), 1876–1892. DOI: 10.1080/09669582.2019.1660670.

Voase, R.N. (ed.) (2002) *Tourism in Western Europe: A Collection of Case Histories*. CAB International, Wallingford, UK.

Williams, A.M. and Shaw, G. (eds) (1999) *Tourism and Economic Development: Western European Experiences*, 3rd edn. Wiley, London.

WTTC (2019) Economic Impact Reports by Country. Available at: https://wttc.org/Research/Economic-Impact (accessed 2 July 2021).

WTTC (2020) *Travel & Tourism Global Economic Impact & Trends 2020*. WTTC, London.

# 1 Informed Developments for a Sustainable Community: An English Case Study in Renewal and Rejuvenation

**Peter Wiltshier***
*Consultant Researcher, Community and Tourism Development NZ at Research Consultancy NZ*

## 1.1 Introduction

This chapter uses various discourses and related recorded case studies to demonstrate the value and purpose of sustainable development goals (SDGs). These discourses reflect the improving capability to utilize local resources and partnerships to underpin an emergent destination management strategy. Tourism can reflect the undeveloped yet innate capacity of a community to adopt a coherent destination management approach to rejuvenation and regeneration that is considered appropriate to the stakeholders and the visitors attracted by the resources and inimitable offer provided.

SDGs can be used to benchmark and optimize the future health of communities from a developed nation perspective. The community at the heart of this research has a specific focus on the usage of existing resources and existing strategies to achieve development commensurate with stakeholders' vision and beliefs. These resources and strategies have been identified in economic growth, innovation in practice, sustainability in community development, and partnership in collaboration between sets of key stakeholders in both community and business development. The chapter also acknowledges the benefit of best practice through experience from a worldwide academy and practice-based background (Norström *et al.*, 2014; Poudel *et al.*, 2016). It uses a case study of a small market town (population 5000) in the English Midlands. The town epitomizes a long-held belief in the positive outcomes of rejuvenation and regeneration in the era of services marketing and against a background of decline in manufacturing and extractive industry that took hold of the local and national economies in the 1980s.

In terms of the research activity, students from tourism courses at undergraduate and postgraduate level were set tasks as part of the revitalization project employing what is often described a real-world-learning (RWL): work-based or problem-based learning techniques to obtain modular credits towards their university awards (Jamal *et al.*, 2011).

The discourse and development process employed to maximize sustainable development planning is predicated on a long-held set of actions emergent from good practice employed over the past 50 years (Wood, 2002). Initial steps to be taken by stakeholders review values, beliefs and preferred shared identity (Galuppo *et al.*, 2014). These steps support the advocacy of

*pwiltshier@yahoo.co.nz

 *Tourism Planning and Development in Western Europe* (eds K. Andriotis *et al.*)
DOI: 10.1079/9781800620797.0001

supportive marketing that can enable adaptive exploration of alternative segments. Establishing key segments that support identity and values requires management and practices to identify achievement of key performance indicators that must be regularly monitored and managed (e.g. Norström *et al.*, 2014). Embedding from the management is a corpus of learning that reflects goals for local stakeholders' education and an ability and capacity established to transform existing practices and, where essential to meet SDGs, a series of resilient alternatives.

These practices and actions, once adopted and embedded in a sustainable plan, can use tourism as part of the learning community. Stakeholders can reflect on visitors' expectations and perception gaps, which can inform specific aspects of sustainable development that reward behaviour changes appropriate for the community (Dangi and Jamal, 2016). This chapter encourages reflection by stakeholders on diverse windows of discourse. The relational perspectives demand that stakeholders respect such perspectives but adopt a methodology to incorporate these variations (Della Lucia and Franch, 2015). The perspective of actor network and structuration additionally is seen as growing new social capital, a sharing economy that understands the diverse needs of actors and which is predicated on partnership and the revisiting of values, beliefs and identity over time (Baggio *et al.*, 2010).

The first part of this chapter identifies the legacy and its content. The second part explores the literature around revitalization, rejuvenation and interpretation of the stories that abound into reasons to be justly proud and future-focused: firstly, future-focused to identify and accrue materials that underpin extant beliefs and values; secondly, to interpret those stories and the lessons learned into materials to underpin destination development in the future. Finally, the chapter outlines lessons learned in the process and recommendations for any destination, especially those more mature places that need skills, capacity and resources to effect change for their future welfare and enduring legacy.

## 1.2 Literature Review

Through the lenses of various discourses and using case studies this chapter explores outcomes for a destination that best reflect the capacity of the community to adopt a coherent destination management approach to rejuvenation and regeneration. These lenses are postmodern, they are reflexive and built around an emergent body of social science research that reflects on the divergent and occasionally contradictory needs of communities facing development dilemmas within a community, business or tourism context (Billington *et al.*, 2007; Napolitano *et al.*, 2007; Beritelli and Bieger, 2014; Sebby and Jordan, 2018; Wise, 2018).

The focus is on the lessons learned from experience, using a case study. Evaluations used in discourse analysis in the social sciences are held up to the lens of inspection. These include relational discourse, structuration, endogeny, stakeholders and structure agency/actor networks. The identities of these communities are not revealed as this is irrelevant. In the face of mounting pressure from central government and against a background of competitive thrust from post-industrial commerce and enterprise, anonymity is best to ensure frank and open discussion (Baggio *et al.*, 2010). Rather than focus on idiosyncratic issues raised by a study of special cases it might be more useful to draw conclusions from discussions which may have universal appeal in terms of solutions for application (some might say best practice) in other locations (Fyall, 2011; Jenkins *et al.*, 2011; Haven-Tang and Jones, 2012).

The components of such lenses of analysis are set out in Table 1.1. In conjunction with data collection and analysis of values, beliefs, identity and current market activity, the discourses permit a socially constructed reflection to accompany a marketing perspective (Norström *et al.*, 2014).

Revitalization and regeneration as concepts and practices are highly dependent on incorporating non-traditional resources and stakeholders in a mix, leading to community development. The proposed model incorporates diverse orientations evolving from heritage and history, environmental concerns from sustainable development, social equity stakeholder views and an array of non-governmental spokespersons (Baggio *et al.*, 2010). Table 1.1 shows a range of stakeholders, their identity within the community and their contribution to the revitalization and regeneration capacity within their community.

**Table 1.1.** Elements and discourses for renewal.

| Discourse | Example | Outcome |
|---|---|---|
| Relational | Local food and beverages | Territorially specific (butchery and milk products) |
| Structuration | Marshalling multiple decision makers | Events specific (festival arts and books) |
| Endogeny | Farmers' market | Farmers/growers |
| Stakeholder | World Heritage Site | Coherent products and services |
| Actor/network | Partnerships and collaborations | Education sponsored by arts and performance |

Adapted from Senge (1990), Saxena (2014), Puhakka *et al.* (2014), Flaccavento (2016), Latour (2004).

There are many areas of the UK that have outstanding stories of the past, borne of legacies of times past that were remarkable for the innovations in those eras that were, at the time, creative, innovative and are now, perceptibly, legendary in their gestation and birth (Bontje and Musterd, 2016; Carley and Spapens, 2017). This reflection is of an era where new forms of work, new industry and creativity were fostered from new knowledge that had the UK, and more specifically certain areas in England, as the centre of that creativity and innovation during the Industrial Revolution (Moore, 2014).

It is the legacy of the Industrial Revolution in the 18th century, an era of unparalleled creativity and innovation, that has caused some despair in key stakeholders and created some dysfunction in more recent centuries – specifically, the years following the Second World War. From a visitor perspective we all need to care more about the way in which 21st-century knowledge economies present services innovation (Wiltshier, 2018). In terms of the 18th-century innovations in heavy industry, the creative and astounding aspects of 200 years ago may now be a little tired and more than a little depressing, both to the visitor and the local (Ashworth and Tunbridge, 2013). Through focusing on discourse, the related corresponding examples and outcomes, a community can identify development processes for necessary and desirable shared outcomes (Table 1.1).

Lessons learned from that industrial legacy provide narratives from the evolving destination story that are captured and reflected in the tales that may be used by stakeholders with support from historians, novelists and diarists, to create an identity, brand and aim for sustainable destination development (see examples from Hartman *et al.*, 2019; Korez-Vide, 2017; Bassano *et al.*, 2019).

The aim is to turn those cultural and heritage remnants that are old, tired and in need of revitalization into the content to steer communities into revitalization as sustainable destinations of distinction for the future (Agarwal, 1999, 2002; Saxena, 2014; Holladay, 2018). Partnerships between the cultural, economic and environmental stakeholders and agents can, and do, embed community values and beliefs easily within the governance and structures necessary to develop destinations for tourism and for community coherence. There are philanthropic goals espoused by stakeholders where the importance of returning something acquired during association with the community leaves expectations, even anticipations, of returning cultural resources and knowledge to the community that drives revitalization (Murzyn-Kupisz, 2012; Aquino *et al.*, 2018; Bec *et al.*, 2019; Little *et al.*, 2020).

In conclusion, the extant literature usefully informs multiple lenses of inspection and highlights discourses that can provide frameworks for a socially balanced way forward, incorporating environmental sustainability goals as well as productivity and community-shared values as targets for future regeneration and revitalization. It is contemporary debate that reassures us that the research is equally focused on triple-bottom line components. These triple-bottom line components include good governance (Nunkoo, 2017), where responsibility as well as rewards are shared by stakeholders. Heritage and culture, seen as creating a sticky cohesive opportunity to share new resources, are based on addressing any social inequality (Calicchia *et al.*, 2017). Inclusivity and sharing of heritage and cultural assets are also highlighted through the discourses (Dragouni and Fouseki, 2018). A community-interest company can also provide resources for expanding the range of cultural resources that provide inimitable competitive advantage and have recently been explored as constituent to a healthy set of achievable SDGs (Fouseki and Nicolau, 2018; Kryder-Reid, 2018).

## 1.3 Methodology

Student-centred, work-based learning was undertaken in June 2017. It was focused on the sustainability of a nascent community interest company (CIC). More specifically, students were engaged to identify and test the development of a marketing, communication and media plan for the destination in collaboration with existing public and private sector stakeholders. To do this, students selected an appropriate framework devised from secondary desk research covering a range of sustainability issues to engage 'best practice' in identifying essential components of marketing and communication for the newly formed CIC.

An initial meeting with business owners/operators was organized by the key tourism business stakeholder and conducted at the local museum by a panel of six students posing questions relevant to tourism development, community development and general business confidence, insights, perceptions and expectations. The student panellists collected varying views clarifying business confidence and expectations from both the public and private sectors. A further cohort of second-year tourism students donned bright orange sweatshirts and conducted primary research with visitors over a week's duration in April and June of 2017 to identify expectations and perceptions. The results of the focus group and consumer survey have been published elsewhere. The overarching findings informed the group working to create the CIC and ongoing action group.

University students arranged a meeting at the local mining museum to discuss and collate opinions and expectations of business stakeholders (owners and operators from a diverse range of service sector businesses). This scoping meeting was chaired by an influential operator. Business owners were invited to participate in the meeting. Representatives of food and beverage outlets, retailers, museums and accommodation provision participated voluntarily. Outcomes included the establishment of a community interest company and a specific focus for collaborative projects with an agreed calendar of events.

Consumers' and residents' opinions were collected over peak and off-peak demand dates. A secondary project was to explore visitor capacity management at sites of sensitivity for the visitor trail around local heritage in a nearby former industrial site that utilized local water, coal, manufacturing clay and textiles. The result was suggestions for good practice carried out elsewhere in (a) marketing and communication and (b) carrying capacity. The additional focus was centred on public and private power sharing. This is a small community with some important natural and manmade features of interest to visitors. Therefore, stakeholders have energized themselves to improve the perception of the community as a destination for further development. The parish council, working with the local authority and key stakeholders engaged in visitor activity, requested support from the university to develop a strategy and structure to take a vision forward for development. Success now depends on the stakeholders working together to make the initial vision into a strategic plan with oversight from key stakeholders (a good mix of public and private shareholders) to entice more visitors and more working together to create a self-sustaining structure for strategy and management over time.

With a strong presence as a destination of distinction for visitors, there is a good case to engage current residents, businesses, local government and visitors in an agreed image, brand and identity based on shared heritage (Alvarez *et al.*, 2016; Della Lucia *et al.*, 2016). A shared heritage is represented in diverse and differentiated services and products which in this case included education, creative arts, film and television, food and retail (a similar project span is viewed holistically in typologies by Saxena, 2014). The initial approach to tourism development can therefore be predicated on this potential agreement for future plans to engage a diverse and enthusiastic market and an engaged and enthusiastic set of actors working with a limited repertoire of supportive individuals (Jeannerat, 2013; Pröbstl-Haider *et al.*, 2014; Wiltshier, 2017). The anticipated outcomes of a diverse and differentiated product offer will depend somewhat on the maturity of the destination, indeed its position in the life cycle, often represented by poor linkages between disciplines and underperforming partnerships (or no partnerships, as observed by Caffyn, 2000). A sequential approach was used to develop a sustainable model that balances resources with identity and brand expectations of the key stakeholders. The initial steps were to undertake an inventory of attractions and amenities to ensure that there was an agreed coherent offer for visitors that accorded with existing infrastructure and visitor resources (see e.g. Jeuring, 2016; Stylidis *et al.*, 2016; Wiltshier, 2017; Fredholm and Olsson, 2018). In the process, skills and resources were identified in stakeholders wishing to expand the tourism sector as well as assigning responsibilities to those stakeholders in forming a community interest company to monitor and manage developments. This is against a background of market forces and neoliberal politics that demands private sector stakeholders take up a protectionist and competitive role from local and central government in the formation of CICs (see, for background, Mosedale, 2016).

## 1.4 Findings

One of the more important outcomes of interventions to provide resources, new knowledge, skills and capacity to communities is the enduring and resonating embedding of new learning within the community. This new learning must necessarily be adopted and enshrined in tourism development, subject to endorsement by stakeholders. This can naturally lead to a course of action with measurable outcomes for the benefit of a wide range of stakeholders whose resources are going to be central to the success of regeneration (Saxena, 2005). Upon reflection with the secretary of the CIC, the following are listed, not in order of importance but perhaps in order of procedural and operational issues within the CIC.

A development association is being formed which represents the wider interests of key stakeholders engaged in business as well as public sector and regulatory activities. Membership is driven by shared business, social and environmental concerns of stakeholders involved in the research with the university: 'The formation of the development association community interest company is the first step in implementing recommendations in the report. The Association will strive to raise funds to implement as much of the report as possible to develop the village for the benefit of the residents, businesses and visitors.'

The association represents the outcomes of the university's work with key community, business and faith stakeholders. Meetings of key stakeholders follow through in the private and public sectors on issues of general and special interest to all interested stakeholders. The association has also tapped into new resources from regional universities and acknowledged the contribution in research and skills currently provided for the community as an outcome of the longitudinal commitment to knowledge sharing and embedding (as demonstrated by Pröbstl-Haider *et al.*, 2014). The achievement of partnerships to achieve sustainable community and business development is more likely assured by the presence of the CIC.

New food outlets and visitor service centres have been created since 2018. There are teething issues with visitor services that concern provision of bathrooms. Ideally these will be reconciled through the CIC as membership of the group comprises representative stakeholders from the encompassing public and private sectors who have responsibility for addressing the expectations of the wider network that includes the residents and taxpayers (as exemplified in Stylidis *et al.*, 2016). An expectation of regeneration of business outlets can engender hope for decent work and economic growth enshrined in aspirations for sustainable development.

Recreational opportunities using heritage gardens and walkways are next on the agenda with an acknowledgement of the adjacent UN World Heritage Site agreement to connect territorially. A 'Heritage Action Zone' is in the process of deployment with significant input from local and regional territorial authorities. This zone comprises heritage listed buildings (conservation zoned), areas of natural beauty and areas acknowledged for recreation. In venturing into heritage development coupled with healthy living, the outcomes for wellbeing are connected to tourism as well as community development.

The CIC has supported initiatives to revitalize and reconstruct heritage events and attractions with support from small- and medium-sized enterprises (SMEs) established in the past decade. There are useful tourism-related examples from the USA (Delconte *et al.*, 2016). The pursuit of RWL is a focus for students (Jamal *et al.*, 2011; Smith, 2016). Students can achieve competency in the study and reporting of complex and messy tourism development and management issues through active participation in projects such as those outlined here. RWL allows students the opportunity to problem solve for organizations outside the university. It permits, within tight timelines, the chance to act as a project worker or project consultant to identify and synthesize solutions and create recommendations at relatively low cost. RWL encourages innovation, creativity, analytical behaviour, knowledge accrual, exchange, storage and retrieval for stakeholders within the community and within education (Guimón, 2013). Education curriculum development is dependent on reflecting, through a consultative process, industry expectations and outcomes from informed discussion with business managers. Education can therefore demonstrate proactive work with industry to reflect the latter's needs and expectations in skill and capacity to engage, solve and produce relevant new knowledge (Guimón, 2013). It is hoped that innovation and creativity through partnership with the university will lead to improved infrastructure, innovation and applications in business that inspire future-focused sustainable business development.

García-Hernández *et al.* (2017) specifically identified the importance of heritage tourism in addressing community conservation and interpretation, which is at the centre of the contemporary renewal and rejuvenation debate through the knowledge transfer process (KTP) in education. This KTP demonstrates how student and educator participation and engagement in community leads to perception of economic as well as social benefits to any community (Dragouni and Fouseki, 2018). In turn, KTP backgrounds and supports campaigns to promote brand and underpin new marketing activity (Smith and Campbell, 2016). Sharing new knowledge helps promote the accrual of social capital for adjacent community tourism development. New knowledge has been freely exchanged through publicly available funding to help develop community-owned assets and (under strict legislation) ensure that emergent ideas are shared through deliberate attempts to use contracted workers or project consultants whose intellectual contributions are no longer privatized and kept strictly available to directly funding public servants (Becker, 2015; Hesse *et al.*, 2015; Tennant *et al.*, 2019).

New social capital is recognized as emergent from the establishment of a CIC (Plow, 2015; Wilson and Desha, 2016). Stakeholders devise various interpretive forms of culture and heritage through narratives and storytellers not engaged with the development process. Storytellers provide the glue that holds this community together; their role is to underpin new branding that is redolent of the values and beliefs that have grown in importance over the past 200 years but may have been neglected as informing brand and image for the 21st century (Corallo *et al.*, 2019). One of the key performance indicators for renovation and rejuvenation is the spread of new stories actively supported by stakeholders that were not traditionally engaged in supply-side branding and identity creation. In this example the CIC has understood the necessity and relevance of this storytelling to interpret a destination and unite stakeholders in a quest to derive social as well as economic capital from tourism (Table 1.2).

The outcome of rejuvenation and the establishment of a CIC to set parameters and conditions for economic, social and environmental benefits to accrue to this community through tourism has been a focus. Quality of life, quality of heritage interpretation and conservation have been paramount for all stakeholders' long-term benefits (Waterton and Watson, 2015; Jelinčić and Mansfeld, 2019). This chapter identifies that, through conservation and protection goals for the community, an outcome in quality of life and a quality of experience for invited guests has been achieved. It is the mix of heritage and performance that underpins the relevance and consistency of the quality of experience for all

**Table 1.2.** The glue and the legacy.

| Discourse | Example | Final outcome of student project |
|---|---|---|
| Relational | Local food and beverages | Developed businesses, business support and mentoring, training and customer service. |
| Structuration | Marshalling multiple decision makers and creation of new management structure | Established a CIC to manage implementation, and measure the impact of the plan |
| Endogeny | Farmers; market | Attracted new higher spending consumers |
| Stakeholder | World Heritage Site | Preserved the historic character of the village.<br>Post photos of recent visit online using hashtags such as #whyilovexxxxxxxx and residents to post stories using #mystorymyvillage |
| Actor/network | Partnerships and collaborations | Created a leading festival and events destination |

Adapted from Senge (1990), Saxena (2014), Puhakka *et al.* (2014), Flaccavento (2016), Latour (2004).

stakeholders. Resources fire up community enthusiasm and shared capital through embodied performativity in the relevant identified community space of heritage and its continued reinvention and making (Crouch, 2017).

A coherent picture of the components of regeneration emerges. This chapter uses discourse, the relevant practice examples and the championed outcomes worked on by staff and students of the university (Table 1.2).

Our initial report identified that existing target markets were justified by the evidence collected at the site. Visitor trends will have an important impact on the future domestic leisure tourism market. Understanding sociodemographic trends is crucial to developing tourism products and services. The growth in intergenerational family holidays (including grandparents and great-grandparents) is already providing growth in the visitor economy and is predicted to have sustainable growth.

The changes in family structures include larger family groups that include children from second marriages. There is a growth in the trend to bring extended families together for short breaks and leisure trips to spend time with family and celebrate big family events such as weddings and birthdays. Sociodemographic trends also include increased ethnic diversity in the population. The leisure tourism needs of ethnic groups are poorly understood in the sector, but this group will become increasingly important. The economic downturn (2008) provided an increase in domestic tourism as consumers stayed at home rather than holiday overseas (as noted in this location by Davison and Ryley, 2016). It is not clear whether this trend will continue and whether consumers' experience and benefits of holidaying at home compensate for the hassle of travelling overseas. An important factor in choosing to holiday at home may be the increasing security risks and uncertainty of travelling overseas. Technology is changing the way consumers access information and make holiday decisions. Technology enables a destination to increase its visibility, reduce its marketing costs and enhance local cooperation in marketing.

The glue that may well help to hold this destination together as a centre for visitor enjoyment, recreation and personal development is based upon shared values, identity reinforced through 300 years of narratives around thermal water and enjoyment of an inland resort. This research identifies, reviews and conceptualizes shared values and beliefs underpinning social capital for the benefit of the majority of stakeholders. Kagan *et al.* (2013) see the embedded learning of 'how to' and 'what is it' as useful social capital as bonding and bridging 'a process through which wellbeing and social capital are both generated and linked together by insight into self and the other' (Kagan *et al.*, 2013, p. 292). However, it is also critical that we view the action taking the strategy forward as in no way dismissive or acknowledging inevitable contradictions and impasses between opposed groups that may not always benefit those involved (Kagan, 2006). As Kagan identifies, such work conducted in the public arena of the community may lead stakeholders to see that 'local people may see the activists as the problem solvers of the community' (2006, p. 1).

A further aspect in the development of a regeneration project that seeks long-term benefits to a variety of dissonant voices is the need for partnership, collaboration and networks that reflect the objective of United Nations' SDGs. Burns (2006) identifies the congruence of outcomes from a variety of multiple partners. The findings reinforce the collaboration between identified key stakeholders that can develop a sustainable tourism offer. The focus is the credible creation of a collaboration process engaging stakeholders' imaginations without burdening those engaged with resource-depleting, process driven activities (Evans, 2016). To undertake this in a strategic way, having identified resources, cooperation between stakeholders, the uplift of existing skills and values and a process to operationalize tourism as a regeneration tool required ongoing analysis of facts (objective) and expectations (subjective) within the framework of the new organization. This needed a conceptual start point in stakeholders and actor/network theory to accommodate the desire to retain at least some competitive advantage within the existing market for tourism.

## 1.5 Conclusion

Through the use of discourse analysis this community can espouse a sustainable destination development process (cf. Dangi and Jamal, 2016). It is also relevant to review findings from other communities that reflect success and wisdom and embedded learning in a variety of lenses from business, education, tourism, services and resources such as Hjalager (1997, 2010) and Ruffin (2010) where collaboration process, product, information and institutional innovation have all been reflected and measurements of a partnership approach partly embedded in the community itself. Implicit is an agreement, a concurrence, among majority stakeholders about the strategic direction of the projects. Cawley and Gillmor (2008) and Bramwell (2006) discuss the importance of 'negotiated power relations' where witnesses seek to concur on the level of agreement on intellectual as well as physical resources at sites.

The CIC has been established with a working committee to manage the implementation, and measure the impact of the sustainable development plan intended to attract new, higher spending consumers while aiming to retain and preserve the historic character of the community.

In evaluation terms, the achievements of the CIC in relation to tangible outcomes such as business support, mentoring, training and customer service has not yet been proven. Unfortunately, the CIC has not yet been able to undertake any funded projects to attract new businesses to the destination. The CIC has as yet not been successful in obtaining a legacy directing the operations and suggestions for initiatives relating to the quality of the visitor experience, nor to the ongoing operations of the coordinated visitor experience as a manager has not yet been appointed. However, better information points have been sought and made available for interpretation and support of the visitor experience and public toilets have been refurbished. In efforts to create a leading festivals and events destination the CIC has been more imaginative and slightly more successful in hosting relevant events to reinforce the brand and identity of the heritage through a heritage trail, sensory trail and a dark tourism initiative with interpretation.

What has been demonstrated through our small-town rejuvenation and regeneration project is the value of new learning in bridging social capital which has also resulted in the establishment of a CIC to take a variety of elements forward. Included in those elements are culture, accessibility, retail offer, heritage and storytelling, better destination marketing and elements emerging of quality of life, reflecting Napolitano *et al.*'s (2007) work in Italy. The destination identity and reinforced values for the stakeholders engaged with the CIC are reflected in the social bonding capital that incorporates capacity to manage change, trust between diverse strands of enterprise and a recognition of the value of exchange (Sebby and Jordan, 2018; Wise, 2018). In addition to articulating the new resources in social capital, more freely exchanged within the community, tangible elements included using key actors, storytellers, performers and artists and their networks in retaining competitive advantage and emergent materials for others to borrow in benchmarking exemplar case studies for the future.

An ACES model incorporating the development knowledge over time, accrual of new datasets and analyses can be seen as relational and structured. This model also requires that development for tourism is in tandem with community goals and these are coherent. Finally, the model leads to long-lasting solutions for ready reference by generations of stakeholders – an enduring legacy that is used by each community – but the model is accessible by a range of communities (e.g. the emergent developing community as much as the mature community). Accrual of new sustainable development actions through tourism needs to be coherent and cohesive to ensure no conflict with other SDG actions (Poudel *et al.*, 2016). We learn from the actions and now plan to store, retrieve and amend data collected and monitoring actions managed by appointed stakeholders in such a manner that education of all stakeholders can be achieved. The outcome is the enduring and transformative effects and benefits of the ACES model on SDGs.

Limitations in this chapter underscore the need to test and re-test management and monitoring of developments conducted by the destination's leaders. Future research must focus on testing the strengths of various discourse elements through deductive empirical work. This research can reduce the bias from vested interests and from the university's syllabus focus on learning rather than practice. This learning inherently brings elements of trust and competency to the fore; community stakeholders, both in the public and the private sectors, deserve to define sustainable development internally rather than import development schedules and lenses from external sources, however well-meaning such intentions may be. Endogenous development requires a high level of trust and engagement at the site to secure skills for ongoing management and monitoring of sustainable local visitor management (see Beritelli and Bieger, 2014; Aquino *et al.*, 2020).

Finally, and importantly, engaging learners from the university proved to be useful in several ways. Academics broadened the appeal of the curriculum through RWL and students benefited from using examples in practice that have supported classroom action. Community stakeholders benefited from the academic input and developed solutions for community development and rejuvenation as well as for tourism expansion.

All stakeholders became acquainted with skills provided for the establishment of a working CIC through both staff and students at the university. The working CIC can develop community and inform the tourism curriculum at the university. The CIC permitted longitudinal research using this case study and is reflected in the discourse analysis employed here. We are now better informed as to what is a learning community and a blueprint for what all stakeholders can do to use cultural heritage resources in various discourses by multiple experiments to provide sustainable competitive advantage to the community through revitalization. Future deductive experiments may be necessary to qualify the balance of discourses and disciplines and achieve coherent long-term goals.

## References

Agarwal, S. (1999) Restructuring and local economic development: implications for seaside resort regeneration in Southwest Britain. *Tourism Management* 20(4), 511–522. DOI: 10.1016/S0261-5177(99)00004-7.

Agarwal, S. (2002) Restructuring seaside tourism: The resort life cycle. *Annals of Tourism Research* 29(1), 25–55.

Alvarez, M.D., Yuksel, A. and Go, F. (eds) (2016) *Heritage Tourism Destinations: Preservation, Communication and Development*. CAB International, Wallingford, UK.

Aquino, R.S., Lück, M. and Schänzel, H.A. (2018) A conceptual framework of tourism social entrepreneurship for sustainable community development. *Journal of Hospitality and Tourism Management* 37(6), 23–32. DOI: 10.1016/j.jhtm.2018.09.001.

Aquino, R., Luck, M. and Schanzel, H. (2020) The outcomes of tourism social entrepreneurship: Host community perspectives. *CAUTHE 2020: 20: 20 Vision: New Perspectives on the Diversity of Hospitality, Tourism and Events*, p. 547.

Ashworth, G. and Tunbridge, J.E. (2013) Heritage and tourism: between practice and theory? In: *Contemporary Issues in Cultural Heritage Tourism*. Routledge, London, pp. 47–59.

Baggio, R., Scott, N. and Cooper, C. (2010) Improving tourism destination governance: a complexity science approach. *Tourism Review* 65(4), 51–60. DOI: 10.1108/16605371011093863.

Bassano, C., Barile, S., Piciocchi, P., Spohrer, J.C., Iandolo, F. *et al.* (2019) Storytelling about places: Tourism marketing in the digital age. *Cities* 87(3), 10–20. DOI: 10.1016/j.cities.2018.12.025.

Bec, A., Moyle, B., Timms, K., Schaffer, V., Skavronskaya, L. *et al.* (2019) Management of immersive heritage tourism experiences: A conceptual model. *Tourism Management* 72, 117–120.

Becker, B. (2015) Public R&D policies and private R&D investment: A survey of the empirical evidence. *Journal of Economic Surveys* 29(5), 917–942.

Beritelli, P. and Bieger, T. (2014) From destination governance to destination leadership – defining and exploring the significance with the help of a systemic perspective. *Tourism Review* 69(1), 25–46. DOI: 10.1108/TR-07-2013-0043.

Billington, R., Cadoppi, V. and Carter, N. (2007) Stakeholder involvement, culture, and accountability in the Blackstone Valley of New England, USA: A work in progress. *Tourism Review International* 11(2), 97–106. DOI: 10.3727/154427207783948919.

Bontje, M. and Musterd, S. (2016) *Inventive city-regions: path dependence and creative knowledge strategies*. Routledge, London.

Bramwell, B. (2006) Actors, power, and discourses of growth limits. *Annals of Tourism Research* 33(4), 957–978.

Burns, P.M. (2006) Innovation, creativity and competitiveness. In: Buhalis, D. and Costa, C. (eds) *Tourism Management Dynamics: Trends, Management and Tools*. Elsevier Butterworth-Heinemann, Oxford, UK, pp. 97–107.

Caffyn, A. (2000) Is there a tourism partnership life cycle. *Tourism Collaboration and Partnerships: Politics, Practice and Sustainability* 2, 203–229.

Calicchia, P., Pitolli, L. and Salonia, P. (2017) The Cultural Heritage Open Laboratory System (CHeLabS). Unfolding a cultural heritage-driven development in science and technology. *Acta IMEKO* 6(3), 4–11.

Carley, M. and Spapens, P. (2017) *Sharing the world: sustainable living and global equity in the 21st century*. Routledge, London.

Cawley, M. and Gillmor, D.A. (2008) Integrated rural tourism: Concepts and practice. *Annals of Tourism Research* 35(2), 316–337.

Corallo, A., Esposito, M., Marra, M. and Pascarelli, C. (2019) Transmedia digital storytelling for cultural heritage visiting enhanced experience. In: *International Conference on Augmented Reality, Virtual Reality and Computer Graphics*, June. Springer, Cham, pp. 221–229.

Crouch, D. (2017) Space, living, atmospheres, affectivities. In: Nieuwenhuis, M. and Crouch, D. (eds) *The Question of Space: Interrogating the Spatial Turn between Disciplines*. Rowman & Littlefield, London, pp. 1–21.

Dangi, T. and Jamal, T. (2016) An integrated approach to "sustainable community-based tourism". *Sustainability* 8(5), 475.

Davison, L. and Ryley, T. (2016) An examination of the role of domestic destinations in satisfying holiday demands. *Journal of Transport Geography* 51, 77–84.

Delconte, J., Kline, C.S. and Scavo, C. (2016) The impacts of local arts agencies on community placemaking and heritage tourism. *Journal of Heritage Tourism* 11(4), 324–335.

Della Lucia, M. and Franch, M. (2015) Case B: Culture-led urban regeneration and brand building in Alpine Italian cities. In: *Harnessing Place Branding Through Cultural Entrepreneurship*. Palgrave Macmillan, London, pp. 122–140.

Della Lucia, M., Trunfio, M. and Go, F.M. (2016) Does the culture of context matter in urban regeneration processes. In: Alvarez, M.D., Go, F.M. and Yuksel, A. (eds) *Heritage Tourism Destinations: Preservation, Communication and Development*. CAB International, Wallingford, UK, pp. 11–21.

Dragouni, M. and Fouseki, K. (2018) Drivers of community participation in heritage tourism planning: an empirical investigation. *Journal of Heritage Tourism* 13(3), 237–256.

Evans, N.G. (2016) Sustainable competitive advantage in tourism organizations: A strategic model applying service dominant logic and tourism's defining characteristics. *Tourism Management Perspectives* 18, 14–25.

Flaccavento, A. (2016) *Building a Healthy Economy from the Bottom Up: Harnessing Real-world Experience for Transformative Change*. University Press of Kentucky, Lexington, Kentucky, USA.

Fouseki, K. and Nicolau, M. (2018) Urban heritage dynamics in 'heritage-led regeneration': Towards a sustainable lifestyles approach. *The Historic Environment: Policy & Practice* 9(3-4), 229–248.

Fredholm, S. and Olsson, K. (2018) Managing the image of the place and the past: contemporary views on place branding and heritage management. *Place Branding and Public Diplomacy* 14(3), 141–151.

Fyall, A. (2011) Destination management: challenges and opportunities. In: Wang, Y. and Pizam, A. (eds) *Destination Marketing and Management: Theories and Applications*. CAB International, Wallingford, UK, pp. 340–359.

Galuppo, L., Gorli, M., Scaratti, G. and Kaneklin, C. (2014) Building social sustainability: multi-stakeholder processes and conflict management. *Social Responsibility Journal* 10(4), 685–701.

García-Hernández, M., de la Calle-Vaquero, M. and Yubero, C. (2017) Cultural heritage and urban tourism: Historic city centres under pressure. *Sustainability* 9(8), 1346.

Guimón, J. (2013) *National policies to attract R&D-intensive FDI in developing countries. Policy Brief, The Innovation Policy Platform*. World Bank, Washington, DC.

Hartman, S., Parra, C. and de Roo, G. (2019) Framing strategic storytelling in the context of transition management to stimulate tourism destination development. *Tourism Management* 75, 90–98.

Haven-Tang, C. and Jones, E. (2012) Local leadership for rural tourism development: A case study of Adventa, Monmouthshire, UK. *Tourism Management Perspectives* 4(1), 28–35. DOI: 10.1016/j.tmp.2012.04.006.

Hesse, B.W., Moser, R.P. and Riley, W.T. (2015) From big data to knowledge in the social sciences. *The Annals of the American Academy of Political and Social Science* 659(1), 16–32. DOI: 10.1177/0002716215570007.

Hjalager, A.M. (1997) Innovation patterns in sustainable tourism: An analytical typology. *Tourism Management* 18(1), 35–41.

Hjalager, A.M. (2010) A review of innovation research in tourism. *Tourism Management* 31(1), 1–12.

Holladay, P.J. (2018) Destination resilience and sustainable tourism development. *Tourism Review International* 22(3), 251–261. DOI: 10.3727/154427218X15369305779029.

Jamal, T., Taillon, J. and Dredge, D. (2011) Sustainable tourism pedagogy and academic-community collaboration: A progressive service-learning approach. *Tourism and Hospitality Research* 11(2), 133–147.

Jeannerat, H. (2013) Staging experience, valuing authenticity: Towards a market perspective on territorial development. *European Urban and Regional Studies* 20(4), 370–384. DOI: 10.1177/0969776412454126.

Jelinčić, D.A. and Mansfeld, Y. (2019) Applying cultural tourism in the revitalisation and enhancement of cultural heritage: An integrative approach. In: *Cultural Urban Heritage*. Springer, Cham, pp. 35–43.

Jenkins, J., Dredge, D. and Taplin, J. (2011) Destination planning and policy: Process and practice. In: Wang, Y. and Pizam, A. (eds) *Destination Marketing and Management: Theories and Applications*. CAB International, Wallingford, UK, pp. 21–38.

Jeuring, J.H. (2016) Discursive contradictions in regional tourism marketing strategies: The case of Fryslân, The Netherlands. *Journal of Destination Marketing & Management* 5(2), 65–75.

Kagan, C. (2006) *Making a Difference: Participation and Well-being*. New Start Publishing, Sheffield, UK.

Kagan, C., Micallef, M., Siddiquee, A., Fatimilehin, I., Hassan, A. *et al.* (2013) Intergenerational work, social capital and wellbeing. *Global Journal Community Psychology Practice* 3(4), 286–293.

Korez-Vide, R. (2017) Storytelling in sustainable tourism management: Challenges and opportunities for Slovenia. *Journal of Advanced Management Science* 5(5), 380–386. DOI: 10.18178/joams.5.5.380-386.

Kryder-Reid, E. (2018) Introduction: tools for a critical heritage. *International Journal of Heritage Studies* 24(7), 691–693. DOI: 10.1080/13527258.2017.1413680.

Latour, B. (2004) On using ANT for studying information systems: A (somewhat) socratic dialogue. In: Avgerou, C., Ciborra, C. and Land, F. (eds) *The Social Study of Information and Communication Technology*. Oxford University Press, Oxford, UK, pp. 62–76.

Little, C., Bec, A., Moyle, B.D. and Patterson, D. (2020) Innovative methods for heritage tourism experiences: creating windows into the past. *Journal of Heritage Tourism* 15(1), 1–13.

Moore, I. (2014) Cultural and Creative Industries concept – a historical perspective. *Procedia-Social and Behavioral Sciences* 110, 738–746.

Mosedale, J. (2016) Conclusion: Tourism and neoliberalism: states, the economy and society. In: Mosedale, J. (ed.) *Neoliberalism and the Political Economy of Tourism*. Ashgate, Farnham, UK, p. 157.

Murzyn-Kupisz, M. (2012) Cultural, economic and social sustainability of heritage tourism: issues and challenges. *Economic and Environmental Studies* 12(2), 113–133.

Napolitano, M.R., De Nisco, A. and Riviezzo, A. (2007) "Urban revitalization and Town Centre Management: the role of Stakeholders". *Proceedings of the 6th International Congress on Marketing Trends*, Organized by ESCP-EAP, Paris, Jan, pp. 26–27.

Norström, A., Dannenberg, A., McCarney, G., Milkoreit, M., Diekert, F. *et al.* (2014) Three necessary conditions for establishing effective Sustainable Development Goals in the Anthropocene. *Ecology and Society* 19(3), 1–8.

Nunkoo, R. (2017) Governance and sustainable tourism: What is the role of trust, power and social capital? *Journal of Destination Marketing & Management* 6(4), 277–285.

Plow, B. (2015) Understanding artists' relationships to urban creative placemaking through documentary storytelling. *Journal of Urban Culture Research* 10, 24–39.

Poudel, S., Nyaupane, G.P. and Budruk, M. (2016) Stakeholders' perspectives of sustainable tourism development: A new approach to measuring outcomes. *Journal of Travel Research* 55(4), 465–480.

Pröbstl-Haider, U., Melzer, V. and Jiricka, A. (2014) Rural tourism opportunities: strategies and requirements for destination leadership in peripheral areas. *Tourism Review* 69(3), 216–228. DOI: 10.1108/TR-06-2013-0038.

Puhakka, R., Cottrell, S.P. and Siikamäki, P. (2014) Sustainability perspectives on Oulanka National Park, Finland: Mixed methods in tourism research. *Journal of Sustainable Tourism* 22(3), 480–505.

Ruffin, F.A. (2010) Collaborative network management for urban revitalization: The business improvement district model. *Public Performance and Management Review* 33(3), 459–487.

Saxena, G. (2005) Relationships, networks and the learning regions: case evidence from the Peak District National Park. *Tourism Management* 26(2), 277–289.

Saxena, G. (2014) Cross-sector regeneration partnership strategies and tourism. *Tourism Planning and Development* 11(1), 86–105. DOI: 10.1080/21568316.2013.839473.

Sebby, A. and Jordan, K. (2018) Rural tourism: a grounded theory on limited stakeholder advocacy. *Archives of Business Research* 6(8), 216–227.

Senge, P. (1990) *The Fifth Discipline*. Currency Doubleday, New York.

Smith, A. (2016) *Experiential Learning*. Edward Elgar, London.

Smith, L. and Campbell, G. (2016) The elephant in the room: heritage, affect and emotion. In: Logan, W., Craith, M.N. and Kockel, U. (eds) *A Companion to Heritage Studies*. Wiley-Blackwell, Hoboken, New Jersey, USA, pp. 443–460.

Stylidis, D., Sit, J. and Biran, A. (2016) An exploratory study of residents' perception of place image: The case of Kavala. *Journal of Travel Research* 55(5), 659–674. DOI: 10.1177/0047287514563163.

Tennant, J.P., Crane, H., Crick, T., Davila, J., Enkhbayar, A. *et al.* (2019) Ten hot topics around scholarly publishing. *Publications* 7(2), 34. DOI: 10.3390/publications7020034.

Waterton, E. and Watson, S. (2015) *The Palgrave Handbook of Contemporary Heritage Research*. Springer, New York, USA.

Wilson, K. and Desha, C. (2016) Engaging in design activism and communicating cultural significance through contemporary heritage storytelling: A case study in Brisbane, Australia. *Journal of Cultural Heritage Management and Sustainable Development* 6(3), 271–286.

Wiltshier, P. (2017) Legacies from nurturers in tourism; inspiring people for communities. *Revista Turismo & Desenvolvimento* 1(27/28), 1993–2007.

Wiltshier, P. (2017) Community engagement and rural tourism enterprise. In: Oriade, A. and Robinson, P. (eds) *Rural tourism and enterprise: management, marketing and sustainability*. CAB International, Wallingford, UK, pp. 53–67.

Wiltshier, P. (2018) New Age visitors and the tourism industry. *International Journal of Religious Tourism and Pilgrimage* 6(3), 49–67.

Wise, N. (2018) Tourism and social regeneration. *Social Sciences* 7(12), 1–6.

Wood, M. (2002) *Ecotourism: Principles, Practices and Policies for Sustainability*. United Nations Environment Programme (UNEP), Paris, France.

# 2 Foreign Tourists Reaching Italy: Transport Mode Choice and Seasonality

**Anna Serena Vergori***
*University of Salento, Lecce, Italy*

## 2.1 Introduction

Mobility decisions characterize all phases of the tourism experience. Transport systems connect tourists from their country of origin to destination regions and are also crucial for inter- and intra-destination movements (e.g. Masiero and Zoltan, 2013). The choice of the transport mode used to reach a destination inevitably impacts the transport mode choice while staying at the destination (Hough and Hassanien, 2010; Aguilò *et al.*, 2012) and poses environmental issues (e.g. Gössling and Peeters, 2007; Peeters *et al.*, 2007). These issues are among the most dealt with by a relevant part of the literature, jointly with the effects of transportation on tourist destinations' development (i.e. Khadaroo and Seetanah, 2008; Pagliara *et al.*, 2015), the impact of specific innovations and the effects of the deregulation of air transport (see among others Graham, 2006; Rey *et al.*, 2011; Pagliara *et al.*, 2015), and the identification of tourist typologies associated with transport modes used at destinations (e.g. Gouveia Barros, 2012).

Although the relationship between tourism and transport has been widely studied from different perspectives, the effects of the transport mode used to reach a destination on the seasonality of tourism demand have received less attention than they deserve. There are some reasons to analyse more deeply the above-mentioned relationship. First, the relevance of the phenomenon of seasonality and the difficulty in coping with it require new strategies. Furthermore, the air transport industry's deregulation from the late 1990s in Europe (Pender and Baum, 2000) impacted travellers' transport mode choices. More competition and the advent of low-cost carriers (LCCs) are the main consequences of the deregulation, resulting in lower airfares. In turn, lower airfares have affected relative prices between air and alternative transport modes, contributing to changes in tourists' travelling habits.

Furthermore, each alternative mode of transport to reach the same destination requires a different travel time, impacting vacation planning throughout the year. Thus, this chapter's underlying hypothesis is that there could be a different seasonal pattern for each transport mode used to reach a destination. This hypothesis allows us to investigate the effects of changed travel habits on intra-year tourism demand distribution. In other words, two main issues are addressed. The first one concerns international tourists' travel habits: are they deeply changed (mainly) because of lower airfares applied by low-cost carriers? In turn, has the possible increase in air transport use contributed to modifying the seasonal pattern of tourism demand? This second issue is whether tourists

*serena.vergori@unisalento.it

 *Tourism Planning and Development in Western Europe* (eds K. Andriotis *et al.*)
DOI: 10.1079/9781800620797.0002

who travel by plane are more or less seasonal than those who travel by other transport modes.

This chapter analyses data concerning international tourism in Italy from January 1997 to December 2019. It confirms that the mode of transport chosen to reach a tourist destination plays a role in determining the seasonal pattern of tourism demand, widening the list of possible seasonality causes. Although the Covid-19 pandemic has temporarily reduced international tourism, the results of our analysis suggest that the tourism sector can play a crucial role in the recovery of the Italian economy, especially if the relationship with transport systems is adequately considered.

## 2.2 Main Issues Related to Seasonality

The seasonality of tourism demand at a destination is an uneven distribution of tourist arrivals in a certain period, characterized by recurrent stability (e.g. Bar-On, 1975; Hartmann, 1986; Butler, 2001). Although it is often analysed on an annual basis (e.g. Lundtorp *et al.*, 2001; Cuccia and Rizzo, 2011; Vergori, 2012; Ruggieri, 2015), other kinds of seasonality – in particular, weekly or daily variations – can be observed (Vergori, 2017). A higher concentration of tourists is observed in the peaks of a season (one or more – two peaks are the most common case) while a lower concentration happens in the rest of the year (off-peak). Sometimes a 'shoulder' season is also observed – that is, between the peak and the off-peak seasons.

This unbalanced distribution of tourists during a year poses many economic, social and environmental issues that have been widely discussed in the literature (Bar-On, 1999; Koenig-Lewis and Bischoff, 2005; Coshall *et al.*, 2015). From an economic point of view, the firms' main problems concern the underuse of their facilities during the off-peak season and their overuse during the peak season. This implies inefficient use of the available facilities. Furthermore, due to higher demand during the peak season, higher prices for goods and services are charged to tourists and residents. In terms of the job market, the main negative effect relates to the creation of temporary employment. Seasonal tourism creates seasonal jobs with well-known consequences on an economy (Krakover, 2000; Baum and Lundtorp, 2001). The main social and environmental problems are related to the congestion of resorts and tourist destinations and the consequences in terms of traffic congestion, waste management, etc.

Identifying seasonality causes is essential to find the proper strategies to manage it according to each destination's needs. The literature has identified many causes (see among others Bar-On, 1975; Butler, 1994; Frechtling, 1996; Butler and Mao, 1997; Baum and Hagen, 1999; Lundtorp *et al.*, 2001). Lundtorp *et al.* provide a taxonomy of the causes identified by the previous literature and distinguish between *push* and *pull* factors. The first group comprises institutional factors, calendar effects, inertia and tradition, social pressure or fashion, and access to the destination. In the second category, climatic conditions, sporting seasons and events are involved. Factors related to both the demand and supply side of the tourism market are considered in the two groups. Thus, seasonality seems to be determined by factors relating to tourists' attitudes and their country of origin (demand-side) – such as inertia and institutionalized holidays – and by attributes of the tourist destination, such as climate, events, provision of services and accessibility. A recent study by Ferrante *et al.* (2018) has highlighted that institutional and sociocultural factors seem to be more causal in determining tourism seasonality in European countries than climatic and environmental factors.

Based on the analysis of the causes of the phenomenon, many strategies to counteract it have been identified. The most cited strategies concern the organization of events during the off-peak season, market and product diversifications, price and tax incentives and structural changes (e.g. Baum and Hagen, 1999; Weaver and Oppermann, 2000; Butler, 2001; Vergori and Arima, 2020). Since seasonality patterns are dynamic and may change over time (Hartmann, 1986; Rosselló Nadal *et al.*, 2004), appropriate policy measures can fit the seasonal pattern of tourism demand with the tourist destination's needs.

Despite intensive efforts by industry and governments, seasonality continues to persist in most tourist destinations, suggesting that the problem is more complex than generally thought. Recently, regarding the determinants of tourism seasonality, Duro and Turrion-Prats (2019) highlighted that 'although researchers may have identified the causes of seasonality [...] greater efforts should be made to establish a more comprehensive theoretical framework. It is also necessary to corroborate this theoretical framework with empirical research that allows one to, among other things, observe the relative strength of each factor and their relative influences' (p. 40).

### 2.2.1 Accessibility and seasonality

The accessibility of a destination is a prerequisite for tourism development. The role of transport infrastructures for the development of tourist destinations has been widely analysed in the literature (i.e. Prideaux, 2000; Khadaroo and Seetanah, 2008; Pagliara *et al.*, 2015).

Efficient transport options enhance the links between tourism-generating regions and tourist destinations. From an economic viewpoint, the less expansive the journey is, the more attractive the destination is. Indeed, the cost of travel is one of the most common explanatory variables used in the estimation of tourism demand (e.g. Loeb, 1982; Smeral, 1988; Crouch, 1994; Lim, 1999; Song and Li, 2008; Cho, 2010). The main literature cites accessibility to a destination among seasonality causes (e.g. Bar-On, 1975; Butler, 1994; Frechtling, 1996; Butler and Mao, 1997; Baum and Hagen, 1999; Lundtorp *et al.*, 2001). However, despite the wide existing literature concerning the relationship between transport and tourism, transportation effects on tourism seasonality have still received too little attention. To the best of our knowledge, there are just a few studies about the impact of LCCs on seasonality (Donzelli, 2010; Graham and Dennis, 2010; Pulina and Cortés-Jiménez, 2010; Chung and Whang, 2011).

These studies deal exclusively with air transport, and they all confirm the positive impact of LCCs on the growth of tourist flows. Regarding the impact on seasonality, they get mixed results. Donzelli's analysis of air traffic flows in three airports of southern Italy highlights a more even intra-year distribution of tourists over time. This result is confirmed by Pulina and Cortés-Jiménez (2010) for Alghero (an Italian tourist destination). In contrast, there seems to be no reduction in seasonality in Korea (Chung and Whang, 2011) or Malta (Graham and Dennis, 2010). However, it is worth noting that these last two studies concern periods of time too short to evaluate the evolution of the seasonal pattern of tourism demand properly.

More empirical research should be performed to understand the causes of seasonal variations of tourism demand related to the accessibility of a destination. In this respect, the analysis of the travel habits of tourists cannot be neglected. Bearing in mind the idea that there could be different seasonal distributions of tourist flows for each transport mode used, it is important to consider tourists' behaviour according to all possible transport modes used to reach a destination.

### 2.2.2 Measuring seasonality

According to the definition of seasonality as an uneven intra-year distribution of tourism flows, it is self-evident that the phenomenon's measurement is based on indicators of inequality. Such indexes tend to be synthetic, and the same value of the index could be associated with different seasonality patterns. Therefore, such indexes should be used jointly with other statistical tools to provide a more accurate analysis.

Following Kuznets (1932), De Cantis *et al.* (2011) propose a methodological framework for analysing seasonal variations in tourism. After plotting the time series against time, highlighting its main characteristics, a further step is the decomposition of the time series into its main non-observable components (trend-cycle, seasonal, residual). The isolation of the seasonal component of the time series is the starting point for further analysis. There are several methods for decomposing a time series (see De Cantis *et al.*, 2011); however, among the most used seasonal adjustment procedures in economic applications we can cite TRAMO-SEATS and X12-ARIMA. The first one is based on a statistical model identified by the TRAMO program, after a pre-adjustment of the time series.

Then, based on the ARIMA model selected, the decomposition of the time series is obtained by the SEATS program. Conversely, X12-ARIMA decomposes the time series by repeatedly applying linear filters that consist of centred moving averages of different lengths. When we focus on the seasonal component, both its pattern and amplitude should be analysed jointly with their persistence or variation. Since similar seasonal amplitude values can derive from different seasonal patterns, seasonal amplitude analysis should be preceded by pattern analysis (Ferrante *et al.*, 2018).

Seasonal patterns can be investigated by seasonal factor indexes, while the amplitude measures comprehend inequality indexes. The Gini index or coefficient (GI/GC) is the most used in the tourism literature. It is sensitive to central observations, that is, to changes occurring during different months with the number of tourists close to the annual average demand, while giving symmetrical weight to extreme observations, that is to months with both the highest and the lowest tourist demand. Theil indexes and the coefficient of variation (CV) are the other two inequality indexes used in the literature. The first is sensitive to the observations in the last positions of the ranking, while the CV is neutral (Duro, 2016). None of them is preferred to the others; each has its peculiarities associated with its own sensitivity; consequently, the choice depends on the research's goals.

## 2.3 Database and Methodology

This chapter uses a database collected by the Bank of Italy. It is based on interviews done with foreigners that take a trip to Italy. The survey is conducted by administering questionnaires at more than 60 international borders (roads, ports, airports and railways). The interviews are taken at the end of the trip. About 130,000 travellers are interviewed every year (see the manual and the questionnaire on the Bank of Italy's website). In the following sections, we analyse data about foreign tourists who visited Italy from January 1997 to December 2019. The analysis focuses on transport modes used by tourists to reach Italy, also considering the tourists' country of origin and the Italian macro-area visited. From March to June 2020, the survey has been suspended due to the Covid-19 pandemic. Data for the year 2020 were published in June 2021, but they give information just for the principal variables. Thus, unfortunately, many data useful for this chapter are not available, such as data about the transport mode used to reach Italy.

We chose the TRAMO-SEATS procedure to isolate the seasonal component of the time series. According to ISTAT (Italian National Institute of Statistics), it is one of the best seasonal adjustment procedures in estimating each non-observable component of a time series. Furthermore, some studies have evaluated the impact of some seasonal adjustment procedures on tourism flow forecasts. More accurate forecasts are obtained when the TRAMO-SEATS procedure, rather than the X-12 ARIMA, has been used to adjust the time series (e.g. Vergori, 2010).

Finally, this chapter proposes the joint use of the GC and the CV as inequality indexes. As we said in Section 2.2.2, GC is the most widely used index in tourism research, and the CV is the neutral one. The Gini ranges between 0 and 1. The higher the value, the higher the inequality distribution; that is, the higher the seasonality.

## 2.4 Foreign Tourists in Italy: Main Characteristics

Italy is the third most important European destination in terms of international tourist arrivals. Since the 1990s the proportion between domestic and foreign tourists in Italy has changed: in 1990 foreign tourists accounted for 35% of total arrivals while in 2019 they represented 49.5% of the total. Since 1990, foreign tourist arrivals have been growing on average at about 4% per year, while domestic tourism has been growing at a 1.8% yearly rate.

The increase in the relative importance of foreign tourists is also due to the deregulation of air transport at the end of the 1990s. The introduction of LCCs has contributed to changing tourists' travel habits, making air transport affordable to many people, not only wealthy people. LCCs' lower fares seem to boost the tendency to divide holidays into sub-periods and take breaks in more than one season (e.g.

**Table 2.1.** Share of travellers according to transport mode and travel reasons (%). (Bank of Italy.)

| | Air | | Railway | | Sea | | Road | |
|---|---|---|---|---|---|---|---|---|
| | 1997 | 2019 | 1997 | 2019 | 1997 | 2019 | 1997 | 2019 |
| Business | 34.8 | 33.5 | 3.4 | 1.5 | 2.6 | 2.0 | 59.2 | 63.1 |
| Tourism | 16.9 | 52.9 | 6.0 | 1.8 | 2.2 | 1.0 | 74.9 | 44.3 |

**Table 2.2.** Foreign tourists by transport mode (%). (Bank of Italy.)

| | 1997 | | | 2019 | | |
|---|---|---|---|---|---|---|
| | Air | Road | Other | Air | Road | Other |
| Germany | 3.9 | 91.8 | 4.3 | 25.8 | 70.6 | 3.6 |
| France | 10.6 | 84.5 | 4.9 | 47.9 | 49.7 | 2.4 |
| Austria | 1.8 | 93.7 | 4.5 | 1.9 | 95.2 | 2.9 |
| UK | 66.1 | 26.7 | 7.1 | 97.1 | 1.8 | 1.1 |
| Switzerland | 4.2 | 75.0 | 20.8 | 8.1 | 85.9 | 6.0 |
| Netherlands | 15.0 | 80.4 | 4.6 | 56.5 | 42.6 | 0.8 |
| Spain | 39.4 | 57.2 | 3.6 | 93.5 | 2.2 | 4.3 |

**Table 2.3.** Percentage of tourists by transport mode according to geographic distribution. (Bank of Italy.)

| | 1997 | | | 2019 | | |
|---|---|---|---|---|---|---|
| | Air | Road | Other | Air | Road | Other |
| North | 7.0 | 86.5 | 6.5 | 28.3 | 68.6 | 3.1 |
| Centre | 43.5 | 43.6 | 12.9 | 85.3 | 12.7 | 2.0 |
| South and islands | 28.3 | 62.1 | 9.6 | 90.3 | 7.3 | 2.4 |

Rosselló Nadal *et al.*, 2004; Lohmann and Duval, 2011). The market's share of LCCs has increased over time. They carried 9.1% of total foreign tourists in 2004 and 54.3% in 2019 (enac.gov.it).

Table 2.1 shows that the travel habits of foreign tourists visiting Italy changed. The use of air transport had increased while all other transport modes had declined in 2019 compared to 1997. It is worth noting that people who travelled for business reasons did not deeply change their travel habits.

The choice of the transport mode used to reach a destination also depends on the distance between the country of origin of tourists and the destination region. From the data analysis, it emerges that the relative importance of the countries of origin changed. In 1997, about 70% of total foreign tourists came from neighbouring countries, while they were about 50% in 2019. In general, foreign tourists who visit Italy come mainly from European countries, and long-haul trips represent 15% of total international arrivals.

Table 2.2 highlights the seven main European countries providing tourists to Italy (in order of importance), according to the transport mode used to reach Italy. The increase in air transport use regardless of the country of origin is clear. It is worth highlighting that the relative importance of tourists coming from France, the UK, the Netherlands and Spain increased in 2019 compared to 1997. Their greater use of air transport has probably contributed to this.

Finally, we have distinguished the Italian destination regions as north, centre, and south and islands (Table 2.3). An increase in the use of air transport emerges in each of these three parts of Italy. However, air transport grew mainly in the centre and the most peripheral

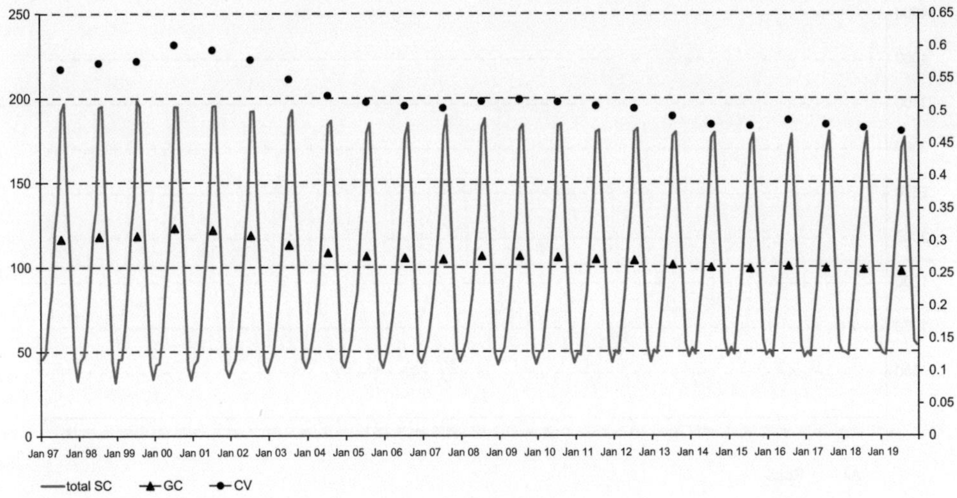


**Fig. 2.1.** Total arrivals: seasonal component and inequality indexes.

parts of the Italian peninsula. Consequently, in 2019 compared to 1997, international tourism grew more in the south and islands (249%) and the centre (114%) than in the north (39%) of Italy. Of course, road transport is the most used in the north of Italy, which is the part closest to the rest of Europe.

## 2.5 TRAMO-SEATS Decomposition

From the previous section, it emerges that tourist arrivals by air and by road represent more than 90% of total foreign tourist arrivals in Italy. Therefore, the analysis hereafter focuses on the time series of tourist arrivals by air and by road. Using the E-Views 8.0 software, we decompose the time series through the TRAMO-SEATS program to distinguishing the trend-cycle component from the seasonal one.

Fig. 2.1 shows the total tourist arrival seasonal component (on the left vertical axis) jointly to GCs and CVs (on the right vertical axis). A one-peak seasonal pattern emerges, with the higher number of tourists concentrated in August, or more generally in the summer season. However, both the visual inspection and the two inequality index patterns highlight a reduction of the seasonal component's amplitude.

To understand the role of transport modes in reducing foreign tourism demand seasonality, Fig. 2.2 plots tourist arrivals by air and by road against time. It confirms the peak in the summer season regardless of the transport mode used to reach Italy. August is the peak month for arrivals by road, while since 2001, July has become the peak month for arrivals by air. Fig. 2.2 confirms the increasing importance of air transport.

Decomposing time series into their main components, one can notice two different trends in tourist arrivals by air and by road (Fig. 2.3). An increasing trend characterizes tourist arrivals by air, while a quite stable trend emerges for tourists who reach Italy by road.

Seasonal components of tourist arrivals by air and by road are shown in Figs 2.4 and 2.5, respectively. Both figures measure the seasonal component on the left vertical axis and the two inequality indexes on the right vertical axis. Comparing the two figures, it emerges that the amplitude of the seasonal component of tourists travelling by road is higher than the one of tourists travelling by air. The GC and CV confirm this; both indexes tend to assume higher values for arrivals by road than by air. In other words, tourist arrivals by road tend to be more seasonal. Furthermore, over time air tourist arrival seasonality has grown, while road arrival seasonality has reduced. It is worth noting that the results of GC and CV corroborate each other and, in both figures, their patterns are cohesive with the seasonal component pattern.

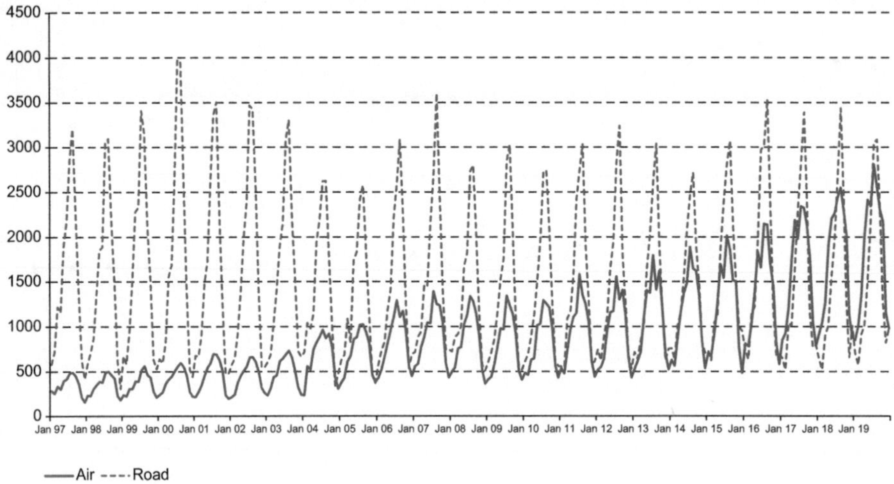


**Fig. 2.2.** Monthly foreign tourist arrivals by air and road.

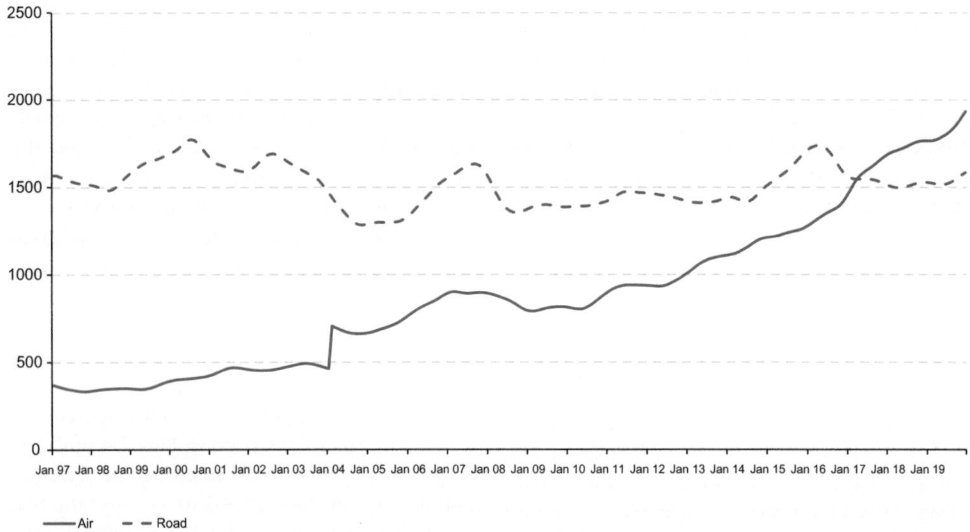


**Fig. 2.3.** Trend component.

## 2.6 Geographic Distribution and Seasonality

Italy is a country with a strong tourist vocation that satisfies a heterogeneous tourism demand. Several kinds of tourism can be enjoyed travelling in Italy, and the same type of tourism shows different aspects according to the peculiarities of each Italian destination. However, Italy is a long and narrow peninsula with two main islands, and the physical access to tourist destinations takes more or less time depending on their geographic positions. Italy borders Europe to the north, while the south and the islands are the most peripheral areas.

This section analyses data disaggregated for the north, the centre, and the south and islands. The aim is to investigate the impact of air transport's increasing use upon the seasonality of the three main Italian macro-areas. The histograms

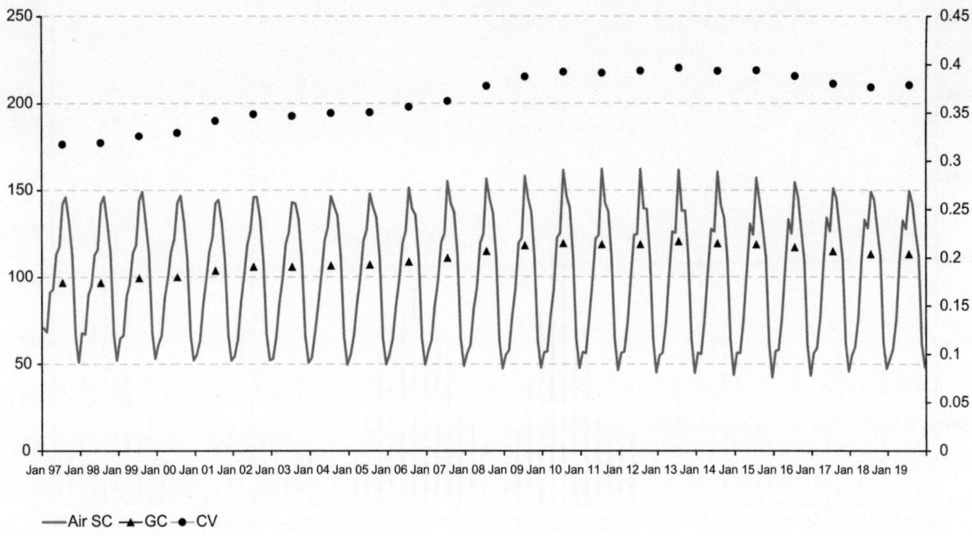


**Fig. 2.4.** Air arrival seasonal component and inequality indexes.

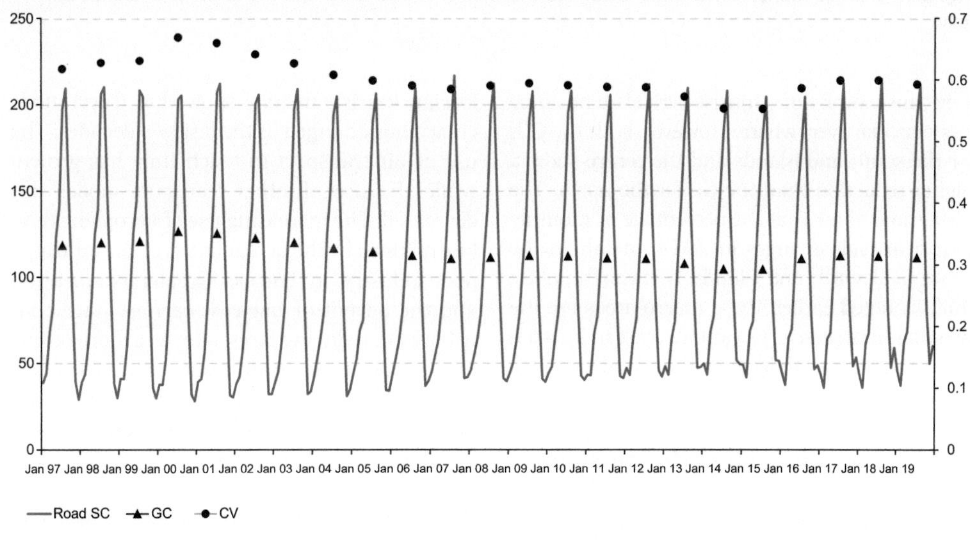


**Fig. 2.5.** Road arrival seasonal component and inequality indexes.

in Fig. 2.6 represent monthly foreign tourist arrivals as a percentage of yearly total arrivals, measured on the left vertical axis. The first two histograms show the monthly share of tourist arrivals in northern Italy in 1997 and 2019. The two histograms in the middle represent the arrivals to central Italy, and the last two histograms are for the south and islands. The GCs are measured on the vertical right axis; they are represented through circles, triangles, and squared indicators, respectively, for the north, the centre, and the south and islands.

A one-peak season emerges in each macro-area in both years. The peak month was August in 1997 while it was July in 2019. In both years, the GC is higher for the south and islands than for the centre and the north, but it is lower in 2019 than in 1997 for all the three parts of Italy. This

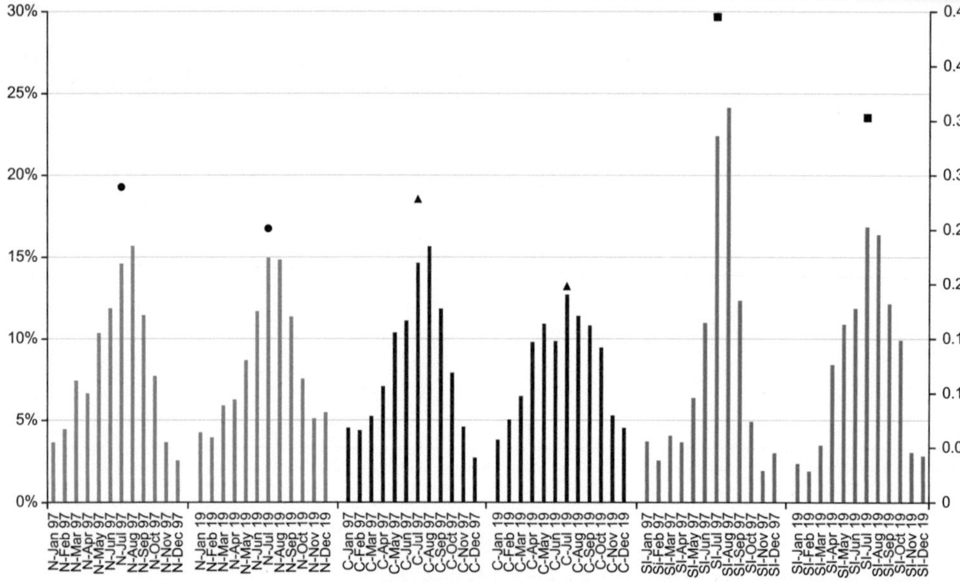

**Fig. 2.6.** Monthly tourist arrivals and GCs – northern Italy, central Italy, and the south and islands of Italy.

is because, over time, tourist arrivals tend to be less seasonal everywhere. However, both the GCs for the south and islands and the centre show a higher reduction than the GC for the north. The histograms show that the percentage of monthly tourist arrivals changes in 2019, mainly in the centre and south and islands. It is worth noting that, as noted earlier, these macro-areas are the two that have seen a relevant increase in international tourist arrivals by air in the period analysed.

## 2.7 Discussion and Conclusions

In the tourism literature, seasonality causes have been widely studied, and they are usually grouped into natural and institutionalized factors. This work contributes to the literature by investigating how the choice of transport mode to reach a destination impacts the seasonal tourism demand pattern. Although some scholars cite accessibility to a destination among seasonality causes, the role of transportation on tourism seasonality has been neglected.

The empirical analysis of Italian data allowed us to answer the questions addressed at the beginning of the chapter. First, data about foreign tourists in Italy show that travel mode choice has changed in the last two decades. The use of air transport to reach Italy has grown, while the use of other transport modes has decreased. The increasing use of aeroplanes can be explained by the deregulation of the air transport industry in Europe. Greater competition and the advent of low-cost carriers have contributed to reducing fares, allowing more people to travel by air. Since air transport allows people to travel long distances in less time, tourists are not bound to travel during the period of the year when they have more free time. This issue allowed us to focus on the second aspect that emerges from our analysis: different seasonal patterns for each transport mode used to reach Italy exist. The intra-year distribution of tourists travelling by air is more even than for those travelling by road. It is worth noting that, although tourist arrivals by road are more seasonal than arrivals by air, in the period analysed, inequality indexes show a decreasing path for arrivals by road and an increasing one for arrivals by air.

The effect on the seasonality of international tourism demand overall is a slight reduction. In other words, the changed travel habits of foreign tourists have contributed to reducing seasonality, and the main reason is a reduction of road use, to be in part replaced by air transport. These results are confirmed by the data analysis disaggregated from the three main Italian macro-areas. A higher reduction of seasonality emerges for both the south and islands and the centre, the two Italian macro-areas that have seen a higher increase in air transport use. Furthermore, a relevant growth in the number of foreign tourist arrivals also emerges for those two areas. This is because competition in the air industry has augmented the number of routes towards peripheral areas.

As in most European countries, tourism in Italy tends to be concentrated in the summer season, which is the most popular period, whatever the transport mode used to reach the country. As we have already said, many factors contribute to determining the shape of the intra-year distribution of tourist flows at a destination; thus, the transport mode choice alone cannot completely change its shape, but it can impact the phenomenon's intensity. This is the main theoretical contribution of this chapter. From a practical perspective, when policymakers aim to reduce seasonal intensity, they should consider the relationship between transport modes and seasonality. More effective results could be obtained, and proper strategies defined according to them. For example, price incentives could encourage air companies to increase the number of flights during the off-peak season.

The results obtained should be corroborated by investigating the same relationship for other countries (along with the availability of compatible data). The analysis of a single country's data is, in fact, the main limitation of this research.

It is worth noting that cars and aeroplanes, which are the most used transport modes to reach Italy, are also the most polluting ones. On the one hand, the increase in air transport use contributes to reducing seasonality in Italian tourist destinations; on the other hand, it pollutes more than other transport modes. Since policymakers are interested in fostering a kind of tourism that is both less seasonal and less polluting, this dual aim deserves more attention. Future research should investigate the relationship between transportation choice and tourism seasonality in the light of environmental issues.

Finally, this chapter has analysed data concerning a period before the Covid-19 pandemic. Data for 2020 give synthetic information about international tourism in Italy. Thus, the impact of transportation on seasonality cannot be estimated for that year. However, it could be valuable here to briefly hint at the effect of the pandemic on international tourism. The proportion of foreign tourists as part of total arrivals in Italy has reduced, from 49.5% in 2019 to about 30% in 2020. Interestingly, the peak in domestic tourism in August 2020 was higher than the peak in previous years. The highest number of foreign tourists (August 2020) is similar to the number of foreign tourists registered during the 2019 off-peak season. Notwithstanding the heavy reduction in the number of foreign tourists, the one-peak seasonality has persisted during the pandemic. In terms of the geographical distribution of tourist flows, the centre is the Italian macro-area with the highest reduction of foreign tourists, followed by the south and the islands. Northern Italy has also had a significant reduction in international tourism but less than the rest of the country.

It is rational to think that the reduction of international tourism is a parenthesis that will be closed with the end of the pandemic. On the contrary, the summer peak is a structural characteristic of tourism in Italy that has been even more evident *during* the pandemic because of reducing infections during summer 2020. Even if data are not available, we can guess that the use of air transport for vacations has been reduced, and at the same time, the use of private vehicles has been augmented. This hypothesis can be corroborated both by the fact that arrivals from the neighbouring countries increased in 2020 and by the fact that the north of Italy is the Italian macro-area with the lower (even if quite high) reduction of international tourism compared to other Italian macro-areas.

The European Council agreed in 2020 to a recovery plan to support the economies of European countries hit by the pandemic. As a result, about €205bn have been allocated to Italy, which has mobilized these funds through the Italian Recovery and Resilience National Plan (RRNP). In addition, the Italian government envisages allocating about €6.68bn to culture and tourism, both considered among the key sectors to enable economic recovery after the pandemic. The Italian government should take the opportunity given by these substantial funds to overcome the tourism sector's weaknesses. In the light of this chapter's results, the European funds should also be spent to promote tourism during the off-peak season, without neglecting that the means of transport chosen to reach a destination plays a role in choosing the period of the year to travel. As we have already said, the aim should be to promote environmentally friendly transport modes that should also be fast and cheap to allow people to move all year round easily. Reaching this goal implies a significant change that will probably take many years. However, the European funds can be a good starting point if they are spent conscientiously.

## References


Aguilò, E., Palmer, T. and Rossellò, J. (2012) Road transport for tourism: evaluating policy measures from consumer profiles. *Tourism Economics* 18(2), 281–293.

Bar-On, R.R. (1975) *Seasonality in tourism. A guide to the analysis of seasonality and trends for policymaking*, Technical series n. 2. The Economist Intelligence Unit, London.

Bar-On, R.R. (1999) The measurement of seasonality and its economic impacts. *Tourism Economics* 5(4), 437–458.

Baum, T. and Hagen, L. (1999) Responses to seasonality: the experiences of peripheral destinations. *International Journal of Tourism Research* 1, 299–312.

Baum, T. and Lundtorp, S. (2001) *Seasonality in Tourism*. Pergamon, Oxford, UK.

Butler, R.W. (1994) Seasonality in tourism: issues and problems. In: Seaton, A. V. (ed.) *Tourism: the state of the art*. Wiley, Chichester, UK, pp. 332–339.

Butler, R.W. (2001) Seasonality in tourism: issues and implications. In: Baum, T. and Lundtorp, S. (eds) *Seasonality in tourism*. Pergamon, Oxford, UK, pp. 5–21.

Butler, R.W. and Mao, B. (1997) Seasonality in tourism: problems and measurement. In: Murphy, P. (ed.) *Quality management in urban tourism*. Wiley, Chichester, UK, pp. 9–23.

Cho, V. (2010) A study of the non-economic determinants in tourism demand. *International Journal of Tourism Research* 12, 307–320.

Chung, J.Y. and Whang, T. (2011) The impact of low-cost carriers on Korean island tourism. *Journal of Transport Geography* 19, 1335–1340.

Coshall, J., Charlesworth, R. and Page, S.J. (2015) Seasonality of overseas tourism demand in Scotland: A regional analysis. *Regional Studies* 49(10), 1603–1620.

Crouch, G. (1994) The study of international tourism demand: a survey of practice. *Journal of Travel Research* 32(4), 41–55.

Cuccia, T. and Rizzo, I. (2011) Tourism seasonality in cultural destinations: empirical evidence from Sicily. *Tourism Management* 32, 589–595.

De Cantis, S., Ferrante, M. and Vaccina, F. (2011) Seasonal pattern and amplitude – a logical framework to analyse seasonality in tourism: an application to bed occupancy in Sicilian hotels. *Tourism Economics* 17(3), 655–675.

Donzelli, M. (2010) The effect of low-cost air transportation on the local economy: evidence from southern Italy. *Journal of Air Transport Management* 16, 121–126.

Duro, J.A. (2016) Seasonality of hotel demand in the main Spanish provinces: measurements and decomposition exercises. *Tourism Management* 52, 52–63.

Duro, J.A. and Turrion-Prats, J. (2019) Tourism seasonality worldwide. *Tourism Management Perspectives* 31, 38–53.

Ferrante, L., Lo Magno, G.L. and De Cantis, S. (2018) Measuring tourism seasonality across European countries. *Tourism Management* 68, 220–235.

Frechtling, D.C. (1996) *Practical Tourism Forecasting*. Butterworth-Heinemann Ltd, Oxford.

Gössling, S. and Peeters, P. (2007) It does not harm the environment! An analysis of industry discourses on tourism, air travel and the environment. *Journal of Sustainable Tourism* 15(4), 402–417.

Gouveia Barros, V. (2012) Transportation choice and tourists' behaviour. *Tourism Economics* 18(3), 519–531. DOI: 10.5367/te.2012.0123.

Graham, A. (2006) Have the major forces driving leisure airline traffic changed? *Journal of Air Transport Management* 12, 14–20.

Graham, A. and Dennis, N. (2010) The impact of low-cost airline operations to Malta. *Journal of Air Transport Management* 16, 127–136.

Hartmann, R. (1986) Tourism, seasonality and social change. *Leisure Studies* 5, 25–33.

Hough, G. and Hassanien, A. (2010) Transport choice behavior of Chinese and Australian tourists in Scotland. *Research in Transportation Economics* 26, 54–65.

Khadaroo, J. and Seetanah, B. (2008) The role of transport infrastructure in international tourism development: a gravity model approach. *Tourism Management* 29(5), 831–840.

Koenig-Lewis, N. and Bischoff, E.E. (2005) Seasonality research: the state of the art. *International Journal of Tourism Research* 7(4–5), 201–219.

Krakover, S. (2000) Partitioning seasonal employment in the hospitality industry. *Tourism Management* 21, 461–471.

Kuznets, S. (1932) Seasonal pattern and seasonal amplitude: measurement of their short-time variations. *Journal of the American Statistical Association* 27(177), 9–20.

Lim, C. (1999) A meta-analytic review of international tourism demand. *Journal of Travel Research* 37, 273–284.

Loeb, P.D. (1982) International travel to the United States: an econometric evaluation. *Annals of Tourism Research* 9(1), 7–20.

Lohmann, G. and Duval, D.T. (2011) Critical aspects of the tourism–transport relationship. *Contemporary Tourism Reviews*, 1–37.

Lundtorp, S., Rassing, C.R. and Wanhill, S. (2001) Off-season is no season: the case of Bornholm. In: Baum, T. and Lundtorp, S. (eds) *Seasonality in Tourism*. Pergamon, Oxford, UK, pp. 89–108.

Masiero, L. and Zoltan, J. (2013) Tourists intra-destination visits and transport mode: a bivariate probit model. *Annals of Tourism Research* 4, 529–546.

Pagliara, F., La Pietra, A., Gomez, J. and Vassallo, J.M. (2015) High speed rail and the tourism market: evidence from the Madrid case study. *Transport Policy* 37, 187–194.

Peeters, P., Szimba, E. and Duijnisveld, M. (2007) Major environmental impacts of European tourist transport. *Journal of Transport Geography* 15, 83–93.

Pender, L. and Baum, T. (2000) Have the frills really left the European airline industry? *International Journal of Tourism Research* 2, 423–436.

Prideaux, B. (2000) The role of transport system in destination development. *Tourism Management* 21, 53–63.

Pulina, M. and Cortés-Jiménez, I. (2010) Have low-cost carriers influenced tourism demand and supply? The case of Alghero, Italy. *Tourism Analysis* 15, 617–635.

Rey, B., Myro, R.L. and Galera, A. (2011) Effect of low-cost airlines on tourism in Spain. A dynamic panel data model. *Journal of Air Transport Management* 17, 163–167.

Rosselló Nadal, J., Riera Font, A. and Sansó Rosselló, A. (2004) The economic determinants of seasonal patterns. *Annals of Tourism Research* 31(3), 697–711.

Ruggieri, G. (2015) Islands tourism seasonality. In: Pechlaner, H. and Smeral, E. (eds) *Tourism and Leisure. Current issues and perspectives of development*. Springer Gabler, Wiesbaden, Germany, pp. 371–383.

Smeral, E. (1988) Tourism demand, economic theory and econometrics: an integrated approach. *Journal of Travel Research* 26(4), 38–43. DOI: 10.1177/004728758802600407.

Song, H. and Li, G. (2008) Tourism demand modeling and forecasting: a review of recent research. *Tourism Management* 29(2), 203–220.

Vergori, A.S. (2010) La stagionalit della domanda di servizi turistici: un'analisi econometrica. *Economia dei servizi turistici* 1, 29–49.

Vergori, A.S. (2012) Forecasting tourism demand: the role of seasonality. *Tourism Economics* 18(5), 915–930.

Vergori, A.S. (2017) Patterns of seasonality and tourism demand forecasting. *Tourism Economics* 23(5), 1011–1027.

Vergori, A.S. and Arima, S. (2020) Cultural and non-cultural tourism: evidence from Italian experience. *Tourism Management* 78.

Weaver, D. and Oppermann, M. (2000) *Tourism Management*. John Wiley and Sons, Milton, Georgia.

# 3 Malta's Tourism Development: Themes, Impacts, Challenges, Patterns and Contrasts: Pointers for a Framework for Short/Long-term Tourism Development

**Marie-Louise Mangion***
*University of Malta, Malta*

## 3.1 Introduction

Islands form part of the set of diverse destinations in Western Europe. Some of those islands, like Malta, are sovereign states, allowing them to independently develop their own economies, manage their own limited resources and govern the commons with a view to economic and social development. Island states, more than other islands, require a financial injection into their economies, which drives governments to prioritize international tourism since it generates income and employment, allows social interaction and addresses insularity. Such prioritization facilitates tourism development. The World Economic Forum's *Travel and Tourism Competitiveness Report 2019* ranks Malta first, globally, in the pillar 'Prioritization of Travel and Tourism'.

Malta's prioritization of tourism, in effect, goes back to the late 1950s. Prior to that, at least from the 18th century and into the 19th century, Malta did register some 'tourism' activity (Cassar, 2020). Throughout the 19th century and for most of the 20th, Malta was a British colony, and consequently a 'fortress economy'. When the British government decided to contract its military base and its expenditure in Malta, an economic planning programme was set out in the *Development Plan for the Maltese Islands 1959–1964* (Malta Government, 1959). It was then that tourism was identified as an economic sector that could support the economy's diversification (Mangion, 2011, 2017), prompting Malta 'to gear up for a full-scale tourism structure' (Cassar, 2020, p. 11). As recognized in the *Master Plan for Malta Tourism* (Italconsult, 1964), Malta, as an archipelago of 316 square kilometres, 'appears well suited to the needs of tourism. Malta is not so small as to be unable to offer a diversity of views and landscapes, nor so large as to lose its sense of being a unified whole' (Italconsult, 1964, p. 1). Since then, tourism has played a central role in Malta's economic and social development.

Fast forward to the span of 11 years prior to the Covid-19 pandemic, that is, 2009–2019. Malta was registering an average annual growth rate in tourist numbers of 7.9%. In 2019, over 2.75 million tourists visited the islands and spent an average of €806 per capita (including flights and accommodation) for an average seven-night stay (NSO, 2020). Tourism accounted for around 12% of gross value added (GVA) at the

*marie-louise.mangion@um.edu.mt

 *Tourism Planning and Development in Western Europe* (eds K. Andriotis *et al.*)
DOI: 10.1079/9781800620797.0003

direct and indirect level, rising to 17% if one accounts for induced effects. To get to that stage, various tourism policies and initiatives were adopted by subsequent governments in cooperation with a series of other actors, foremost being the private sector involved in tourism. Debates surrounding tourism policy decisions have touched on, among other matters, tourism's economic importance versus its vulnerability; the call for continuous growth even amid concerns about the islands' capacity to absorb even more tourists; the balance between economic requirements and social and environmental impacts; and the host community's hospitality versus the resident community's tolerance to increased tourism activity. These are themes which surround and underlie, in effect, most debates on tourism development elsewhere too, but they take on an added edge and dimension in the context of a small island state, which is dependent on tourism activity for its economic survival and characterized by a high population density.

This chapter considers further this circumstancing for tourism in Malta, setting it against the backdrop of Maltese governments' decisions across the years as articulated in national tourism policy documents dated 1959 to 2021. It thereby explores tourism development from a policy perspective. It focuses on the policy content of 15 official documents, identifying first concepts and themes underlying Malta's tourism development, secondly tourism's impacts and challenges as recognized in the policy documents and then drawing parallels and contrasts which feature therein. Based on this analysis, a framework relating to short-term and long-term tourism development is proposed. The specific research questions may, accordingly, be framed as follows:

- What concepts and themes relating to tourism development emerge from policy documents?
- What tourism impacts and challenges are recognized in Malta's tourism policy documents?
- What patterns and contrasts can be drawn relating to the course of Malta's tourism development?
- What supports short-/long-term tourism development?

This chapter responds to these questions, allowing also for comparative analysis across similar destinations. It is worth noting further, in preamble, that it is because Malta has a history of documented tourism policy of over 60 years, because it is so small and impacts are easily observed – to the extent that they are evident and tangible – that this destination offers a good starting point and case study to answer these questions, and hence, possibly, yield conclusions that might have broader applicability and scope.

## 3.2 Tourism Development: Island Destinations and Malta

The understanding of the concept of development evolved from that limited to economic growth to a more encompassing definition incorporating political, social, cultural and environmental aspects (Ruggieri, 2016). For all the extensive relevant literature, consensus on a precise meaning for 'tourism development' has not as yet been reached (Hall, 2000; Saarinen *et al.*, 2017). Two dimensions of tourism development may be considered: one is the way tourism progresses over time; the other is the more recent perspective of tourism development being associated with wellbeing, both at a societal and individual level (Sharpley, 2020). Through planning and policy, it can be argued, destinations can 'revitalise their tourism sector and sustain its viability' (Andriotis *et al.*, 2019, p. 2). Tourism planning is thereby perceivable as a strong tool 'for guiding tourism to a development path that creates benefits and wellbeing beyond the industry and its core operations' (Saarinen *et al.*, 2017, pp. 308–309).

Tourism development on islands is influenced by the particular and distinct attractiveness arising from geographic detachment and from the appeal afforded by the sea or waters surrounding them. This physical detachment, though, poses some challenges through reduced accessibility, which in turn often acts as an attraction in itself. This binarism of opportunities and challenges, and consequent positive and negative impacts, appears to be pertinent for small island destinations.

Drawing on Parra-López and Martínez-González's (2018) literature review on island destinations, economic, environmental and social impacts can be characterized as follows:

- Tourism's economic contribution, for islands, provides much-needed injection in terms of earnings and increased tax revenues. Given its nature and its tendency to have a higher multiplier than some other economic activities, tourism creates linkages with various sectors, utilizing a broader labour force. The downside of this, however, is that tourism activity on islands tends to expand at the expense of local economic activities, such as agriculture and fishing, though these in themselves can act as attractions and support tourism in return, creating a symbiotic relationship. Some islands tend to have size constraints and limited economies of scale, but tourism can help in overcoming these through expanding market size. Tourism helps to improve transport connectivity and communications given its dependence on this infrastructure, but it may also lead to high transport costs and increases in product prices. Land, often a limited resource in smaller islands, may experience rapid price increases, accompanied by speculation. Resources may be shifted more towards infrastructural development required to support tourism than to that which is needed by resident communities. Tourism is often characterized by seasonal demand and it is through destination strategies that a more even spread can be achieved.
- Increased human activity on islands, brought by tourism, places augmented demand on environmental resources, which in return may benefit from increased protection and valuation. Tourism can lead to nature protection and biodiversity conservation as awareness and appreciation of the island's natural environment increases. On the other hand, waste disposal, pollution, biodiversity loss, land take-up and urban sprawl at the expense of rural areas, are among tourism's negative environmental impacts.
- Tourism, from the sociocultural perspective, benefits communities through improved living standards and through the consolidation of culture and traditions. Tourism's negative sociocultural impacts primarily are a consequence of the increased numbers of tourists beyond unsustainable levels.

Most of these impacts are reflected in literature examining Malta's tourism, including works by Boissevain (1977, 1979, 1996); Lockhart (1997); Mangion and Vella (2000); Blake *et al.* (2003); Pollacco (2003); Dodds (2007); Briguglio (2008); Chapman and Speake (2011); Jones (2015); Mangion (2017, 2020a, b); and Cassar and Avellino (2020). What emerges from this scholarly literature about Malta's tourism is that tourism contributes significantly to economic development, with clear benefits to the local population; yet negative impacts accrue when tourism activity is not managed effectively to the detriment of that same constituency. Economic benefits and tourism's importance for the economy can often sway the local population toward tolerance of negative impacts. This trade-off can be rendered more balanced and less detrimental through policy decisions and choices relating to spatial planning and development permits, to mention two obvious points of strategy that can be powerful and influential in a small island context.

One of the reasons for tourism's negative impacts is, according to Baldacchino and Milne (2000), the 'actions or omissions of the agents linked to tourism in the destinations' (Parra-López and Martínez-González, 2018, p. 135). Public agents involved in tourism are many, as they include not only authorities responsible for tourism but also those responsible for culture, environment, education, health, infrastructure, transport and many other policy fields. It is in the nature of a national tourism policy that it will reflect, and be a function of, initiatives and actions – and omissions and oversights (in all senses of this latter word) – of diverse public authorities. Through policy documents, public agents communicate about issues they recognize and the direction they intend to adopt. It is in awareness of that fact that this chapter, through content analysis of tourism policy documents, uses the lens of policy to understand public agents' perspective and decisions relating to Malta's tourism development, its impacts and challenges. Based on this literature review, Fig. 3.1 depicts a theoretical framework for understanding what influences

**Fig. 3.1.** Theoretical framework for tourism development in island destinations. (Author's compilation based on a review of scholarly literature.)

tourism development in island destinations. The next section explains the methodology adopted for this research, outlining how the content analysis was carried out.

## 3.3 Methodology

The methodology adopted for this research was guided by reference works on content analysis, including those by Krippendorf (2018) and Neurendorf (2019). Content analysis of Malta's development plans and tourism policy documents dating back to 1959 was carried out (Table 3.1). Content analysis is used to determine the presence of themes and concepts within the text and draw inferences (Krippendorf, 2018). This supports the purpose of understanding Malta's tourism development by identifying patterns within the themes and concepts that emerge from the selected documents' text (Neurendorf, 2019) by allowing data to be condensed to concepts that shed light on the phenomenon being researched 'by creating categories, concepts, a model, conceptual system, or conceptual map' (Elo *et al.*, 2014, p. 1). As explained by Ercan and Marsh (2016), historically content analysis has been focused on 'counting the number of mentions of specific items' (p. 316); it has now evolved and is 'viewed as a method for interpreting qualitatively generated data' (p. 316). The distinction lies in that 'Quantitative analyses typically rely on the soundness of a priori measurement instruments; qualitative and critical analyses usually rely on the expertise of an expert scholar. In quantitative content analysis, the empirical process is independent of the particular scholar; in qualitative or critical message analyses, it is not' (Neurendorf, 2019, p. 10). (This reliance on the 'expertise of an expert scholar' was not a concern in this study, given the author's research and her work experience as a practitioner in the field of tourism research, strategy and policy within Malta's tourism authorities.) The dichotomy between qualitative and quantitative content analysis is not quite favoured by Krippendorf given that ultimately both focus on text.

The process adopted in conducting this content analysis started by selecting the relevant documentation to be analysed. Criteria were developed to guide the selection of the documentation, namely:

**Table 3.1.** List of official policy documents. (Author's compilation.)

| |
|---|
| *Development Plan for the Maltese Islands 1959–1964* |
| *Master Plan for Malta Tourism* 1963 |
| *Development Plan for the Maltese Islands 1964–1969* |
| *Third National Development Plan* 1969–1974 |
| *National Development Plan for Malta 1973–1980* |
| *Il-Pjan ta' Żvilupp fi Kliem il-Poplu, 1981–1985* |
| The Maltese Islands Tourism Development Plan 1989–2005 |
| *Carrying Capacity Assessment for Tourism in the Maltese Islands – Tourism Development Policies 2000–2010* |
| Malta Tourism *Authority Strategic Plan 2000–2002* |
| Malta Tourism *Authority Strategic Plan 2002–2004* |
| *Malta Tourism Authority Strategic Plan 2004–2006* |
| *Tourism Policy for the Maltese Islands 2007–2011* |
| *Tourism Policy for the Maltese Islands 2012–2016* |
| *National Tourism Policy 2015–2020* |
| *Malta Tourism Strategy 2021–2030 – Recover, Rethink, Revitalise – Stakeholders and Public Consultation* (undergoing public consultation at the time of writing) |

- the inclusion of official, national-level policy documents that relate to tourism planning and/or development, in part or entirely;
- the exclusion of strategies or plans related to a particular location, area or niche market.

Fifteen national policy documents were identified, as listed in Table 3.1, remarkably spanning from 1959 to 2021, and defining successive governments' direction for tourism.

Adopting an inductive approach, this research was carried out through conventional content analysis that allows the codes to emerge from the texts, as explained by Hsieh and Shannon (2005). An integral part of the procedure of content analysis is the coding of text, which involves working through the entire document, selecting text segments of relevance to the topic being researched and assigning evaluative codes to that text. The reviewed literature, synthesized in Fig. 3.1, broadly informed the selection of the relevant text relating to tourism development from the policy documents. Given the inductive approach adopted for this research, concepts and themes were allowed to emerge and were explored (as will be presented in the next section). For the purposes of this exercise, each policy document was read thoroughly and coded in chronological order to facilitate the identification of how policy thinking about Malta's tourism development evolved. Coding was done by the author but internal reliability was ensured through intracoder checks. Codes were then sorted into categories based on the relationships and links of the codes.

The findings presented here place an emphasis on the results that emerge from the more qualitative nature of the content analysis, defined as 'the subjective interpretation of the content of text data through the systematic classification process of coding and identifying themes or patterns' (Hsieh and Shannon, 2005, p. 1278).

## 3.4 Results: Malta's Tourism Development – Concepts, Themes, Impacts and Challenges

Walton (2009) and Butler (2015) agree with Saarinen *et al.* (2017) that 'it is crucial to understand the historical nature and context of tourism development and planning' (p. 311). Mangion (2017) mapped out Malta's historical development of tourism from 1958 to 2015, outlining the country's policies for tourism and the performance over those 57 years. Seven development stages were identified together with seven dominant policy modes, reflective of 'the existence of a relationship between policy (policy output) and tourism performance (outcomes)' (Mangion, 2017, p. 205).

By digging deeper into each tourism policy document, through content analysis and following categorization of the codes, 20 concepts relating to tourism development were identified, as outlined here. Table 3.2 presents the concepts portrayed in each policy document. Concepts are listed chronologically in the order each concept was identified.

### 3.4.1 Concepts: a comparison across policy documents

Nine concepts, as shown in Table 3.2, were identified from the *Development Plan for the Maltese Islands, 1959–1964*, which set the path for future tourism development on the concept of tourism growth and an economic focus. Until then, tourism activity was not 'large enough to amount to a really substantial tourist industry' (Malta Government, 1959, p. 37) in spite of 'Malta's climate and historical interest hav[ing] always attracted visitors' (1959, p. 37). Tourism was to be triggered by focusing on the tangible offer, doing 'everything [...] to realise Malta's assets of its sun and sea, and its historic associations and buildings, to build up a steady tourist trade, which will bring direct and immediate benefit to the islands' economy' (1959, p. 37). The policy recognized that tourism development required both government intervention as well as private sector investment. The government's role involved setting up the Government Tourist Board with 'executive powers to carry out a programme of tourist development' (1959, p. 37). The private sector was encouraged to invest in hotels. Government provided capital grants and tax concessions, in the hope that it would 'encourage the expansion of first-class hotel accommodation, a prerequisite for any substantial increase in the tourist trade' (1959, p. 37). Public spending, then described as a 'very substantial provision' (1959, p. 37), was allocated. The importance of accessibility for island tourism was also recognized with reference being made to the daily services by air to the Continent and the UK, to passenger steamers and the occasional cruise liner. This development plan echoed most of the influential factors depicted in Fig. 3.1. However, trade-offs, environmental and social considerations, and size constraints were not mentioned. On the other hand, the *Master Plan for Malta Tourism* (1963) was insightful. It emphasized economic development and tourism growth, but introduced new concepts: 'the country's maximum tourist capacity' (Italconsult, 1964, p. 57) based on an average density index, spatial planning, tourist zoning, area planning and urbanistic regulation – all matters that are still debated today.

The *Development Plan 1964–1969* (Malta Government, 1964), building on the first plan, also aimed at tourism growth and focused on economic benefits. Tourism development was directed to develop the island into a first-class tourist resort, attracting a higher spending tourist – a reference to upmarket tourism. Reference to environmental resources was minimal and, where present, limited to their potential to generate economic prosperity. The *Third National Development Plan 1969–1974* (Malta Government, 1969) pushed for further growth through diversification into non-UK markets and introduced the concept of addressing seasonality.

A shift in policy was adopted in the *National Development Plan for Malta 1973–1980* (Malta Government, 1973), which regarded tourism as a vulnerable economic activity, but targeted mass tourism. Air Malta, the national airline, was set up, recognizing the need for connectivity. Grants programmes were halted and channelled towards industrial activity. *Il-Pjan ta' Żvilupp 1981–1985* (Malta Government, 1981) continued shifting public resources towards manufacturing, while utilizing price competitiveness against other Mediterranean destinations. Tourism was still viewed as vulnerable and risky. The risk of tourism's negative social impact was just mentioned through a short phrase. In spite of the potential risks, no mitigating measures, such as urban planning, were adopted. The resulting negative consequences are still felt today.

*The Maltese Islands Tourism Development Plan* (1989–2005) (UNDP and WTO, 1989) emphasized the value of tourism planning,

**Table 3.2.** Concepts relating to tourism development in Malta.

| | Tourism growth | Economic focus | Public spending | Government role | Governance | Focus on tangible offer | Private sector | Marketing | Connectivity | Environment | Spatial planning | Carrying capacity | Upmarket | Seasonality | Sociocultural elements | Competitiveness | People in industry | EU Membership | Sustainability | Distribution channels |
|---|---|---|---|---|---|---|---|---|---|---|---|---|---|---|---|---|---|---|---|---|
| *Development Plan for the Maltese Islands 1959–1964* | X | X | X | X | X | X | X | X | X | | | | | | | | | | | |
| *Master Plan for Malta Tourism* (1964) | X | X | X | X | X | X | | | | X | X | X | | | | | | | | |
| *Development Plan for the Maltese Islands 1964–1969* | X | X | X | X | X | X | X | X | X | X | X | | X | | | | | | | |
| *Third National Development Plan 1969–1974* | X | X | | | | | | X | | | | | | X | | | | | | |
| *National Development Plan for Malta 1973–1980* | X | X | X | | X | | | | X | | | | | | | | | | | |
| *Il-Pjan ta' Żvilupp fi Kliem il-Poplu 1981–1985* | X | X | | X | X | X | | X | X | X | | | | X | X | X | | | | |
| *The Maltese Islands Tourism Development Plan 1989–2005* | X | X | | X | X | X | X | X | | X | X | | X | X | X | | X | X | | |
| *Carrying Capacity Assessment for Tourism in the Maltese Islands – Tourism Development Policies* (2001) | X | X | | X | X | X | X | | | X | | X | | | X | | | | X | |
| *MTA Strategic Plan 2000–2002* | X | X | | X | X | X | X | X | X | X | X | X | | X | | | X | X | X | X |
| *MTA Strategic Plan 2002–2004* | X | X | | | | | | X | | X | | X | X | X | | | X | X | X | X |
| *MTA Strategic Plan 2004–2006* | X | X | | | | X | | | | X | | X | X | X | | | X | X | X | X |

*Continued*

**Table 3.2.** Continued

| | Tourism growth | Economic focus | Public spending | Government role | Governance | Focus on tangible offer | Private sector | Marketing | Connectivity | Environment | Spatial planning | Carrying capacity | Upmarket | Seasonality | Sociocultural elements | Competitiveness | People in industry | EU Membership | Sustainability | Distribution channels |
|---|---|---|---|---|---|---|---|---|---|---|---|---|---|---|---|---|---|---|---|---|
| *Tourism policy for the Maltese Islands 2007–2012* | X | X | X | X | X | X | X | X | X | X | X | X | X | X | X | X | X | X | X | X |
| *Tourism policy for the Maltese Islands 2012–2016* | X | X | | | X | X | X | X | | X | X | X | X | X | X | X | X | X | X | X |
| *National Tourism Policy 2015–2020* | X | X | | | X | X | X | X | X | | | | X | X | | X | X | | X | X |
| *Malta Tourism Strategy 2021–2030* | X | X | | | X | X | X | X | X | X | | X | X | X | | X | X | X | X | X |

Note: Shaded cells indicate introduction, emphasis, redirection or evolution of the concept. (Author's compilation based on content analysis of tourism policy documents.)

steering tourism development to a different level. It recognized that 'little will be achieved without the leadership and guidance of the public sector and the trust and responsiveness of the private sector' (1989, p. 5), prompting aspects of governance. Government actions were focused on product and tourist quality, seasonality and market diversification, redesigning marketing plans, shifting dependence away from the traditional sun and sea holiday market and targeting middle-/high-class tourists. Accession to the European Community (today's EU) was mentioned and considered a positive boost to tourism. The absence of land use policy and of a strategic structure plan were highlighted as detrimental to the country, which was missing a mechanism to integrate and control development to the detriment of the environment. This plan incorporated, as shown in Table 3.2, a number of concepts which were given a fleeting mention in previous plans, but it was in this document that detailed planning of those concepts and evident recognition of issues and potential solutions were presented.

Malta's limits to tourism growth were identified in the *Carrying Capacity Assessment* [CCA] *for Tourism in the Maltese Islands* (Mangion, 2001) following almost three years of consultation and research. Analysis of the spatial–ecological environment, the sociocultural environment and the economic–political environment led to proposals for tourism development policies. Elements within the private sector were not, then, too accepting of the CCA report's proposed limited growth scenario for tourism development, while other stakeholders thought it was not sufficiently daring. Nonetheless, subsequent Malta Tourism Authority (MTA) strategic plans and the national tourism policy documents adopted the CCA proposals and the concept of sustainable development, a leap from previous documents. The MTA's mission was to 'advance the economic and social activity of tourism in the national interest by working with all stakeholders to develop a sustainable industry for current and future generations' (Malta Tourism Authority, 2000, p. 20).

*The Tourism Policy for the Maltese Islands 2007–2011* (Ministry for Tourism and Culture, 2006) brought together many issues and concepts referred to in earlier documents, but added on sharper focus on the macroeconomic environment, governance, competitiveness and sustainability – concepts which were echoed by subsequent policy documents until 2021. Malta's tourism entered a new phase as international market trends were tapped: independent travel, online booking systems and low-cost airlines led to a route development policy which was actively implemented. Connectivity to facilitate accessibility and a wider spread of distribution channels were the concepts driving tourism development.

Concurrently, sustainable development principles started to guide decisions. Environmental and social considerations were given more attention in the 2006 document than in previous policy documents. The 2006 document states that 'Environment protection is one way of sustaining an economic activity and a living for many locals who depend on the tourism sector. It is our policy to move more in this direction and ensure that no tourism development results in the destruction and upsetting of sensitive ecological habitats as well as the degradation of historical–cultural resources' (2006, p. 39). It also recognizes that the environment is 'an important element in the authentic tourism experience' (2006, p. 45). Similarly, the 2012–2016 policy (Ministry for Tourism, Culture and the Environment, 2012) expressed its support for 'national policies and initiatives which ensure a high level of environmental conservation and awareness as well as those which maintain Malta's cultural, social and environmental resources' (p. 67). These mark a higher level of commitment to achieving environmental objectives.

*The Tourism Policy for the Maltese Islands 2012–2016* policy went one step further, setting specific environmental and social goals (besides economic ones), promoting sustainable practices and directing decisions accordingly, reflecting the title of its foreword: 'Sustainable Tourism Growth for Malta and Gozo'. This policy sought to drive tourism development to achieve a stronger competitive edge, providing higher value added across the entire value chain by attracting a broader spread of niche markets. The *National Tourism Policy 2015–2020* (Ministry for Tourism, 2015) echoed much of this along with the previously recognized concepts of competitiveness, sustainability and air route network. The *Malta Tourism Strategy*

*2021–2030 – Recover, Rethink, Revitalise* (Ministry for Tourism and Consumer Protection and Malta Tourism Authority, 2021) once again echoed the call for sustainability. In the context of the pandemic, economic benefits, tourism earnings and the sector's profitability are emphasized, while recognizing the challenges in achieving tourism development in a context where the country's general appearance needed improvement, investment in bedstock required redirecting to other key areas and hotel rates were spiralling downwards. The pandemic years, 2020 and 2021 (and possibly beyond that), provided destinations with the opportunity to pause and think about the nature of tourism development. Assessing past policy documents is a good start for reflecting on the future of tourism development.

### 3.4.2 Themes underlying Malta's tourism development

From these 20 concepts, as shown in Fig. 3.2, five main themes emerge as underlying Malta's tourism development, namely (1) growth; (2) national prosperity; (3) governance; (4) tourism offer; and (5) market orientation. Each helps to lend energy and momentum in the development of tourism. Each has developed and evolved over time as a result of policy choices and in reflection of international trends not only in the industry itself but also in the relevant thinking within the various related fields and areas, allowing for initiative and innovation – at least at the policy formulation stage and on paper. Each theme is present in *each* of the policy documents, though with different emphasis and perspectives, as explained here.

#### *Growth*

Growth is central in all policy documents. It is still top of the political agenda, which equates growth with increases in the number of tourists visiting the islands. The question 'How much can Malta's tourism grow?' was not raised during the first 40 years of the country's tourism development, except in the 1963 *Master Plan* which introduced the idea that there is a limit, stating that 'the maximum number of foreign tourists that can be adequately supported by the coastal strip is 10,000' (Italconsult, 1964, p. 57). This question and the issue of capacity was raised again by Jeremy Boissevain and Peter Serracino Inglott (an anthropologist and philosopher, respectively) who stressed that 'Research is urgently needed to help establish the optimum level of tourism this densely populated resort can safely absorb' (1976, p. i).

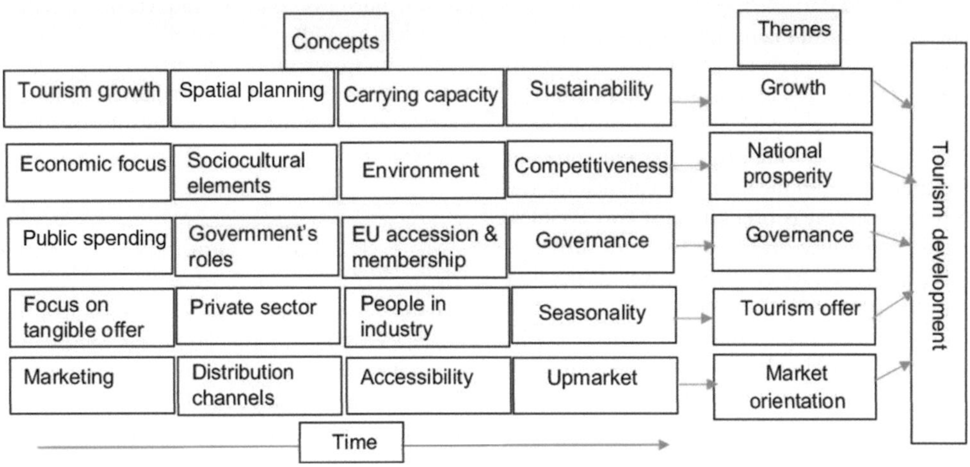


**Fig. 3.2.** Themes underlying tourism development in Malta. (Author's compilation based on content analysis.)

Consequences of growth on land use were only recognized in the 1963 and 1989 documents wherein spatial planning was referred to. (In 1990, Malta issued a first draft of the *Structure Plan for the Maltese Islands* regulating land use.) It was only in the late 1990s that the question relating to tourism growth and capacity made it to the political agenda, triggering the *Carrying Capacity Assessment for Tourism in the Maltese Islands* (Mangion, 2001). The content analysis indicates that this leveraged a policy shift (articulated, not necessarily implemented) towards sustainability, which was adopted by subsequent policy documents.

#### *National prosperity*

Four concepts point to the theme 'national prosperity', as shown in Fig. 3.2. Traditionally, prosperity was defined only in terms of economics. It is increasingly recognized that national prosperity incorporates social and environmental dimensions. In parallel, recent literature on national competitiveness and the current work of the World Economic Forum, among others, link competitiveness to this broader definition of prosperity. Malta's tourism policies also reflected this evolution in thinking about national prosperity.

#### *Governance*

Fig. 3.2 shows four concepts grouped under the theme 'governance'. Over the years, government played a major role in Malta's tourism development as a policymaker, a regulator, a facilitator for setting the right environment for tourism, a financier and a partner for the private sector. It is over the past two decades, incidentally since the Malta Tourism Authority (MTA), with its broader powers, was set up, that aspects, such as integration of policies, coordination, structural reform, improved public consultation and stakeholder involvement, have taken more prominence in policy documents, with the *Tourism Policy for the Maltese Islands 2012–2016* dedicating a whole section to 'The Right Governance Approach' and specific policy responses. Since 2004, when Malta joined the EU, supranational governance, implying policy transfer, was also evident.

#### *Tourism offer*

Four concepts merge under the theme 'tourism offer' (Fig. 3.2). Malta's tourism policymakers focused on Malta's tangible offer, placing emphasis on the beach offer and accommodation facilities. As time passed and Malta's tourism activity matured, attention was devoted to other elements in the tourism value chain, with culture and urban heritage being given more attention and financing once Malta joined the EU. Policy documents recognized the private sector's role, but one can note changes over time in the relationship between the public sector and the private sector. A similar change is noted in how tourism policy documents cogitate about people working in the industry. In addition, policies sought to address seasonality, with the 1989 *Master Plan* placing this as a key strategy that has been retained ever since.

#### *Market orientation*

Tourism development in destinations such as Malta is often determined by international market trends as long as the right environment for tourism activity to take place is provided. Tourism institutions, over the years, marketed Malta as a tourist destination tapping a range of geographic and niche markets, including cultural, diving, English language learning, meetings, incentives, conferences and exhibitions (MICE), business tourism and visiting friends and relatives (VFR). The content analysis shows that Malta's tourism development is dependent on connectivity by air and requires a mix of distribution channels reflective of international market trends.

### 3.4.3 Impacts

Tourism's economic, social and environmental impacts outlined in the literature review on island destinations feature to different extents in these policy documents. Certainly, it is the positive economic impacts of income injection and employment generation which are mostly recognized. Initial plans projected expected economic returns, whereas recent policies have emphasized tourism's economic contribution. Negative economic impacts are not quite dealt with in the documents. High transport costs do not feature, maybe because most public transport is subsidized. Land take-up, speculation, high land and property prices are common in Malta, but policy documents do not associate this with tourism activity. Admittedly, the cause is a complex web of factors.

Most of the policy documents limitedly acknowledged tourism's environmental impacts. The *Master Plan for Malta Tourism* (1964) observed that 'due regard has not always been given to environmental requirements' (Italconsult, 1964, p. 6), whereas *The Maltese Islands Tourism Development Plan* (1989) admitted that Malta faced 'major problems associated with conservation and preservation' (p. 124), while calling for an improved planning process. It is chiefly *The Tourism Policy for the Maltese Islands 2012–2016* which acknowledges these environmental impacts, adopting policies to control and mitigate these.

Recognition of positive sociocultural impacts focuses on increased living standards, conservation and promotion of cultural heritage and traditions. Over time and as tourist volumes increased, policy documents acknowledged tourism's social impacts, notably after the CCA study was conducted.

### 3.4.4 Challenges

Managing tourism in an island state, characterized by a small land area, a high population density and limited resources, poses particular challenges. The content analysis showed that the commonly acknowledged challenges included: attracting a higher spending and more discerning tourist; increasing real earnings per capita; the risk of economic overdependence; and tourism's vulnerability and viability. As the existing stock of accommodation increased, challenges related to land use, prices spiralling down and excess supply of bedstock were recognized. As Malta experienced tourism growth, demand-instigated challenges included managing visitor flows. Recently, as international market trends changed, identifying the optimal mix and share of dominant players in the market became a challenge. With Malta's level of tourism capital and businesses' expectations of profitability, the threat of unrestricted and unlimited growth increased, making the shift to a more sustainable model of tourism development more challenging. Linked to that is the unarticulated but implied challenge of governance and stewardship of the commons.

The pandemic has added a further layer of challenges as, increasingly, power struggles among stakeholders and conflicting interests influence policy choices. At this critical juncture, determining the country's future tourism development path is the main challenge. That can be informed by insights on emerging patterns and contrasts.

### 3.4.5 Patterns and contrasts

Reflecting on the concepts, themes, impacts and challenges outlined above, a number of patterns and contrasts can be drawn. It is evident that some would also be applicable to a non-island destination. However, within the context of a small island destination with a high population density, as in Malta's case, these take on a more acute meaning as the challenge grows more problematic due to limitations concerning size and resources.

#### *Predominance of growth and economic dependence*

Growth and economic prosperity remain top of the agenda. The impact of this is that tourism is recognized as important and is prioritized. On the other hand, the Maltese economy, in spite of further diversification into other service sectors, is still tourism dependent, rendering policymakers cautious about strategies that may limit growth, in spite of acknowledging boundaries imposed by society, the environment and land space.

#### *Spatial planning*

Land use is one of the major challenges the country has, not only because of its small land area and high population density but also because of the various (often conflicting) demands it is called upon to meet as a country. Spatial planning is indispensable in a small island like Malta. A tension often arises between the original scope of a policy and its implementation. Much can be said about tourism development, spatial planning and permitting in the context of Malta, but that would be material for a different study.

### *Public and private sector interdependence*

The content analysis portrays that, since 1959, government steered tourism development in a direction to which the private sector responded. There have also been scenarios, such as that relating to the introduction of low-cost airlines, where representatives of the private sector lobbied government to adopt decisions favouring this prop for tourism development. Government, following discussions and assessments, did adopt that prop albeit with some caution initially. Collaboration between government and the private sector is necessary for tourism development, but both parties must understand their different roles and responsibilities.

### *Dependence on government subsidization*

Malta's tourism development has been dependent on public funding. With the aim of stimulating the industry, subsequent governments provided grants or loans to tourism service providers, subsidies and/or marketing support to tour operators or to airlines. This puts into question whether the sector would be viable without such public funding. However, what one needs to bear in mind is the long-run impact of subsidization: in the short run a subsidy boosts the sector, then as time passes production of tourism falls, though compared to the situation prior to the subsidy being implemented, it remains at a higher level (Schubert and Brida, 2008; Mangion, 2011).

### *Priority to accommodation*

Malta's tourism development has been based on the provision of accommodation given that it is essential in the tourism value chain. Investment, employment generation, linkages with other sectors and the potential to tap different markets were declared reasons for its prioritization. Over the years, accommodation establishments were beneficiaries of grants, interest-free loans, public land contracts and permitting, but also suffered downturns at particular periods. At the end of 2019, Malta had 39,487 beds in serviced accommodation and 16,140 beds in licensed self-catering accommodation. The 2021 strategy indicates that it is envisaged that 'all currently licensed bed-stock plus beds in the pipeline [would amount to] circa 100,000 beds/36.5 million available bed-nights…' (p. 8). These figures imply that Malta would have 316 beds per square kilometre, in an already densely populated country, and 100,000 beds by 2030 means that the current licensed bedstock will almost double in ten years and create bedstock equivalent to that created in the past 60 years. This betrays that the pressure for and the threat of unlimited and unconstrained growth is present, deeply contrasting with all messaging and rhetoric relating to sustainability, environmental respect and minimizing negative social impacts.

The level of bedstock influences the path of tourism development in that excess supply of beds and low room and occupancy rates push market players and policymakers to attract more and more tourism demand. As demand starts increasing and rates rise, entrepreneurs consider reinvesting, creating further pressure on land and other resources, as recognized in the 2021 strategy. The pattern repeats itself.

### *'Few (but) good beaches' and cultural heritage*

In spite of having few good beaches on the island, Malta for many years focused its efforts on developing tourism around the coast, rendering short-term gains. Malta was stronger in terms of the concentration of cultural heritage assets. These would have required proper conservation and presentation prior to being marketed for tourism purposes, a process which was taken up properly, through EU funding, only relatively recently.

### *Dependent on international market trends and air travel*

Malta's tourism development is market-oriented and, understandably, heavily dependent on the island being accessible. Growth was achieved when tourism authorities actively tapped international market trends: mass tourism, tour operator and package travel in the 1970s, 1980s and 1990s; independent travel, online booking systems and low-cost airlines in the new millennium.

#### *The commons and competitiveness*

The commons, such as the skyline, the landscape, natural habitats and biodiversity, are only valued for what they give or have the potential to give to the tourism business. Otherwise, they are exploited or viewed in this light. This is further compounded by other economic activities such as speculative construction which, for example, is dramatically changing the skyline of the islands and hence the character of the archipelago. Increasingly, the commons will play a more important role in the country's attractiveness, in residents' quality of life, in how Malta develops its tourism offer and in tourists' desired experience. Ultimately, Malta's destination competitiveness will be affected (Mangion, 2020a).

#### *Contrasts*

A number of paradoxes can be spotted in Malta's tourism development. Each is related to at least one theme, as explained below.

- Growth: the paradox of success indicates that though unintended consequences accrue, growth is considered a positive.
- National prosperity: prosperity and competitiveness depend on economic, social and environmental aspects but the latter two require further attention.
- Governance: policy leadership is evident while the private sector spurs government. Policy intent is clear but a policy implementation gap exists.
- Tourism offer: profitability is sought but often that profitability is dependent on some form of government subsidization.
- Market orientation: while numerous efforts are made to tap the international market, the shift to the upmarket segment remains a challenge.

### 3.4.6 Framework for short-term and long-term development

The results of the content analysis have signalled particular patterns and paradoxes in Malta's tourism development, providing pointers to short-term and long-term tourism development as depicted in Fig. 3.3. This proposed framework for short/long-term tourism development is based on Malta's case study, but it may resonate with other destinations. It assumes a supportive institutional structure.

It is clear that Malta's tourism development has been driven by the needs set by the economic context, often seeking to address the crises faced at the time. The risk involved here is that short-term considerations would take priority over any potential long-term implications.

Short-term tourism development demands relatively quick results in response to policy decisions. Some of Malta's approaches, as listed in Fig. 3.3, allowed tourism activity to grow, though of course that comes with some negative impacts and trade-offs too, which have not always been recognized, mitigated, addressed or managed. Each of these approaches has its consequences and potential risks, which would need to be assessed prior to any policy decision being taken for short-term results. This decision in itself is political and would therefore also be influenced by the political context at the time.

Long-term tourism development is often seen as the sum of previous short-term approaches. However, sustainability and competitiveness require a further political commitment to long-term considerations and planning. Fig. 3.3 outlines how each of the short-term factors can be rethought and redesigned to steer Malta's tourism development for the long term, even in the context of a pandemic.

## 3.5 Discussion and Managerial Implications

The 15 tourism policy documents are evidence of Malta's political commitment and prioritization of tourism, elements that have supported tourism development for over 60 years. It is probable that such prioritization will remain in the years to come. Yet, that prioritization needs

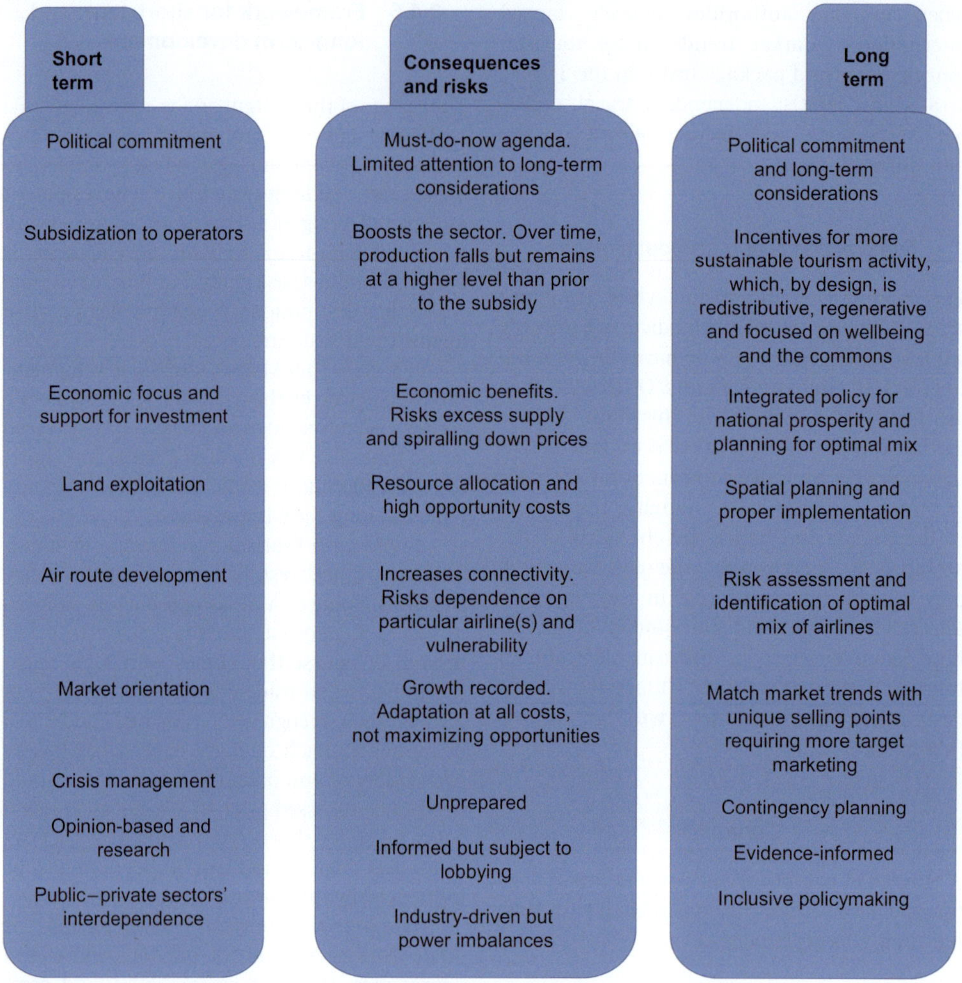


**Fig. 3.3.** Short- and long-term tourism development in Malta. (Author's compilation.)

to be accompanied by refined consideration of potential future impacts.

The content analysis exposed that issues that were of concern for tourism development back in the early days of tourism are still significant policy issues today, though the extent or nature of the problem has evolved. This implies that issues may be addressed in the short term but they may still resurface in the long term, especially if they are of a systemic character that is not easily countered.

The manner in which issues are dealt with often influences or sets the path for future decisions. Most policies echoed concepts introduced in previous policy documents, reflecting the incremental approach to policymaking. At particular points, however, policy shifts were identifiable. The managerial implication of this is that today's policymakers and tourism practitioners need to be aware of their influence on future policies.

Evidently, tourism has strengthened and supported Malta's economy and created jobs. Much capital and investment is tied up in tourism enterprises. In this scenario, policymakers may not be willing to risk adopting a different strategy to tourism development, spiralling into a vicious circle for further growth, even though that may be counterproductive for tourism itself. The Covid-19 pandemic has brought an abrupt halt to this, causing a 76.1% decline in tourism activity in 2020 when compared to 2019 (NSO, 2021), and highlighting tourism's vulnerability and defenceless as aeroplanes have been grounded and social interaction, which is what tourism is all about, kept minimal. Taking decisions about the next steps on the path of tourism development will be influenced by that stock of capital which requires critical mass to operate profitably. If government were to seriously consider redirecting tourism to more sustainable levels of activity, the matter of existing tourism capital may constrain the policymaker unless directional action is taken. Ultimately, the return from tourism should be higher than the overall cost of tourism.

Given all this, policies and plans, including spatial and economic plans, should be redistributive and regenerative by design, even more so in the case of island destinations such as Malta, where such approaches can prove to be more powerful and influential. Undeniably, this approach will involve competing interests and power struggles. Governance structures and systems can be strengthened to reset that balance of power and ensure implementation. In this regard, three critical points seem to be missing from Malta's policy documents:

1. The role of and cooperation with the third sector. Few mentions of the third sector were made in any of the policy documents (three times in each of the 2007 and 2012 policies; single mentions in the 2015 and 2021 documents). Intersectoral cooperation is even more important in an island destination with limited resources but having people who are passionate about particular experiences that tourists may be willing to visit and even possibly pay for.
2. Residents' voice in the process of tourism policy and planning. This is a requirement for achieving sustainability, but, in spite of consultation processes, it is somewhat absent. One reason for this could be that residents are aware of tourism's economic importance and employment spread, making them tolerate a higher level of tourism's negative impacts. This is true at a national policy level. The scenario somewhat changes at a more localized level when applications for planning permission for tourism projects such as hotels or other amenities are submitted. Recently, residents and their respective local council(s) have vociferously opposed such tourism projects, particularly large-scale ones.
3. Distribution of income. This was reflected in some of the earlier tourism policy documents and particularly in those of 2007 (12 mentions) and 2012 (four mentions). However, this appears to be missed in recent policy documents. This finding suggests that tourism development is directed for the benefit of tourism businesses and tourists rather than for, or at least, inclusive of, that of residents.

All three points parallel Moscardo's (2011) findings about tourism planning models wherein destination residents play a minor role. However, if sustainability and competitiveness are to be at the core of tourism development, then the wellbeing of Malta's residents should be central in future tourism policies. It will involve rethinking the possibility of shifting from tourism development which places growth at its centre to tourism development which puts human wellbeing at its core.

## 3.6 Conclusion

This chapter has brought together the two dimensions of tourism development, i.e. time and the perspective of wellbeing. It explored, from the lens of policy and through content analysis of 15 policy documents spanning over 60 years, concepts, themes, impacts and challenges. These pointed to particular patterns and contrasts in Malta's tourism development. Informed by the findings and analysis, a framework for short-term and long-term tourism development was formulated.

Malta's case study provides empirical evidence to demonstrate that growth, national prosperity, governance, the tourism offer and market orientation underlie the island's tourism development. It confirms the theoretical knowledge relating to tourism development, but adds that political commitment and prioritization, national prosperity, policy leadership, spatial planning, subsidization and a mix of distribution channels critically influence tourism development in a context such as Malta's. In addition, due to policy incrementalism, the history of tourism development itself influences future tourism development. Malta's tourism development raises the question as to whether tourism development should assume growth or be redefined.

Malta's tourism development supports Weaver's claim that 'The proposed constituent paradox, however, is that out of both necessity and endowment, small islands are also sustainable sites of impressive resilience and innovation capable of providing peak experiences that foster positive tourism centrality within contexts of optimal autonomy and retained cultural and environmental distinctiveness' (Weaver, 2017, p. 19). Viewing Malta's tourism development from the policy perspective uncovered public agents' thread of thinking about the prioritization of tourism. Therein lies a power struggle between the short-term model of tourism development, which seeks to quench the thirst of existing and planned stock of capital with further tourist numbers, and the long-term model for tourism development, which promotes a more redistributive and regenerative approach. This year, 2021, provides those public agents with an opportunity to choose the future course for Malta's tourism development, and certainly this would be true of their equivalents in many other destinations. It is hoped that within such scope as there ever realistically is for the outcome, longer-term foresight and vision can prevail over short-termism and its consequences.

## References

Andriotis, K., Stylidis, D. and Weidenfeld, A. (eds) (2019) *Tourism Policy and Planning Implementation – Issues and Challenges*. Routledge, Abingdon, UK.

Baldacchino, G.Y. and Milne, D. (2000) *Lessons from the Political Economy of Small Islands: The Resourcefulness of Jurisdiction*. Macmillan, Basingstoke, UK.

Blake, A., Sugiyarto, G. and Sinclair, M.T. (2003) *The Economic Impact of Tourism in Malta – Report for the Malta Tourism Authority*. University of Nottingham, Nottingham, UK.

Boissevain, J. (1977) Tourism and development in Malta. *Development and Change* 8(4), 523–538. DOI: 10.1111/j.1467-7660.1977.tb00754.x.

Boissevain, J. (1979) The impact of tourism on a dependent island: Gozo, Malta. *Annals of Tourism Research* 6(1), 76–90. DOI: 10.1016/0160-7383(79)90096-3.

Boissevain, J. (1996) But we live here: perspectives on cultural tourism in Malta. In: Briguglio, L., Archer, B., Jafari, J. and Wall, G. (eds) *Sustainable tourism in islands and small states: case studies*. Pinter Publishers, London, UK, pp. 220–240.

Boissevain, J. and Serracino Inglott, P. (1976) *Tourism in Malta*. University of Malta, Malta.

Briguglio, L. (2008) Sustainable tourism in small island jurisdictions with special reference to Malta. *Journal of Tourism Research* 1(1), 29–39.

Butler, R. (2015) The evolution of tourism and tourism research. *Tourism Recreation Research* 40(1), 16–27. DOI: 10.1080/02508281.2015.1007632.

Cassar, G. (2020) Tourism and Malta: a narrative of resilience, pragmatism and survival. In: Cassar, G. and Avellino, A. (eds) *Tourism and the Maltese Islands – Observations, Reflections and Proposals*. Kite Group and Institute for Tourism, Travel and Culture, University of Malta, Malta, pp. 1–20.

Cassar, G. and Avellino, M. (2020) *Tourism and the Maltese Islands – Observations, Reflections and Proposals*. Kite Group and Institute for Tourism, Travel and Culture, University of Malta, Malta.

Chapman, A. and Speake, J. (2011) Regeneration in a mass-tourism resort: The changing fortunes of Buġibba, Malta. *Tourism Management* 32(3), 482–491. DOI: 10.1016/j.tourman.2010.03.016.

Dodds, R. (2007) Malta's tourism policy: standing still or advancing towards sustainability? *Island Studies Journal* 2(1), 47–66. DOI: 10.24043/isj.199.

Elo, S., Kääriäinen, M., Kanste, O., Pölkki, T. and Utriainen, K. (2014) Qualitative content analysis: a focus on trustworthiness. *SAGE Open* 1–10.

Ercan, S.A. and Marsh, D. (2016) Qualitative methods in political science. In: Keman, H. and Woldendorp, J.J. (eds) *Handbook of Research Methods and Applications in Political Science*. Elgar, Glos, UK, pp. 309–322.

Hall, C.M. (2000) *Tourism Planning: Policies, Processes and Relationships*. Pearson, Harlow, UK.

Hsieh, H.-F. and Shannon, S.E. (2005) Three approaches to qualitative content analysis. *Qualitative Health Research* 15(9), 1277–1288. DOI: 10.1177/1049732305276687.

Italconsult (1964) *Master Plan for Malta Tourism*. Italconsult, Rome, Italy.

Jones, A. (2015) Economic and labour market implications of climate change on the tourism sector of the Maltese Islands. *Xjenza Online* 3, 128–140.

Krippendorf, K. (2018) *Content Analysis: An Introduction to the Methodology*, 4th edn. Sage, London, UK.

Lockhart, D.G. (1997) 'We promise you a warm welcome': tourism to Malta since the 1960s. *GeoJournal* 41(2), 145–152. DOI: 10.1023/A:1006884808270.

Malta Government (1959) *Development Plan for the Maltese Islands 1959–1964*. Department of Information, Valletta, Malta.

Malta Government (1964) *Development Plan for the Maltese Islands 1964–1969*. Valletta, Malta.

Malta Government (1969) *Development Plan for the Maltese Islands: Third National Development Plan 1969–1974*. Valletta.

Malta Government (1973) *National Development Plan for Malta 1973–1980*. Office of the Prime Minister, Valletta, Malta.

Malta Government (1981) *Il-Pjan ta' Żvilupp fi Kliem il-Poplu – Malta fis-snin 1981–1985 (Malta: Guidelines for Progress, 1981–1985)*. Economic Division, Office of the Prime Minister, Valletta, Malta.

Malta Tourism Authority (2000) *Malta Tourism Authority Strategic Plan 2000–2002*. Malta Tourism Authority, Valletta, Malta.

Malta Tourism Authority (2002) *Malta Tourism Authority Strategic Plan 2002–2004*. Malta Tourism Authority, Valletta, Malta.

Malta Tourism Authority (2004) *Malta Tourism Authority Strategic Plan 2004–2006*. Malta Tourism Authority, Valletta, Malta.

Mangion, M.L. (2001) *Carrying Capacity Assessment for Tourism in the Maltese Islands – Tourism Development Policies 2000–2010*. Ministry of Tourism, Valletta, Malta.

Mangion, M.L. (2011) Evidence-based Policy-making: Achieving Destination Competitiveness in Malta. PhD thesis. The University of Nottingham, Nottingham, UK.

Mangion, M.L. (2017) Tourism in Malta: Policies and Performance (1958–2015). In: Vassallo, M.T. and Tabone, C. (eds) *Public Life in Malta II – Papers on Governance, Politics and Public Affairs in the EU's Smallest Member State*. Department of Public Policy, University of Malta, Msida, Malta, pp. 175–211. Available at: https://www.um.edu.mt/library/oar/handle/123456789/52243

Mangion, M.L. (2020a) The historical future of tourism: The case of Malta's policies. In: Yeoman, I. and McMahon-Beattie (eds) *The Future Past of Tourism – Historical Perspectives and Future Evolutions*. Channel View Publications, Bristol, UK, pp. 39–52.

Mangion, M.L. (2020b) Malta's destination competitiveness and policy considerations: an initial analysis. In: Cassar, G. and Avellino, A. (eds) *Tourism and the Maltese Islands – Observations, Reflections and Proposals*. Kite Group and Institute for Tourism, Travel and Culture, University of Malta, Malta, pp. 105–124.

Mangion, M.L. and Vella, L. (2000) *The Economic Impact of Tourism in Malta*. Malta Tourism Authority, Valletta, Malta.

Ministry for Tourism and Consumer Protection and Malta Tourism Authority (2021) *Malta Tourism Strategy 2021–2030 Recover, Rethink, Revitalise – Stakeholders and Public Consultation*. Malta Tourism Authority, Valletta, Malta. Available at: https://issuu.com/visitmalta/docs/maltatourismstrategy_2030_v7

Ministry for Tourism and Culture (2006) *The Tourism Policy for the Maltese Islands 2007–2011*. Ministry for Tourism and Culture, Valletta, Malta.

Ministry for Tourism, Culture and the Environment (2012) *The Tourism Policy for the Maltese Islands 2012–2016*. Ministry for Tourism, Culture and the Environment, Valletta, Malta.

Ministry for Tourism (2015) *National Tourism Policy 2015–2020*. Ministry for Tourism, Valletta, Malta.

Moscardo, G. (2011) Exploring social representations of tourism planning: issues for governance. *Journal of Sustainable Tourism* 19(4–5), 423–436. DOI: 10.1080/09669582.2011.558625.

National Statistics Office (2020) News Release Inbound Tourism: December 2019. Available at: https://nso.gov.mt/en/News_Releases/Documents/2020/02/News2020_017.pdf (accessed 12 November 2020).

National Statistics Office (2021) News Release Inbound Tourism: December 2020. Available at: https://nso.gov.mt/en/News_Releases/Documents/2021/02/News2021_018.pdf (accessed 4 February 2021).

Neurendorf, K.A. (2019) *The Content Analysis Guidebook*. Sage Publications, Thousand Oaks, California.

Parra-López, E. and Martínez-González, J.A. (2018) Tourism research on island destinations: a review. *Tourism Review of AIEST – International Association of Scientific Experts in Tourism* 73(2), 133–155. DOI: 10.1108/TR-03-2017-0039.

Pollacco, J. (2003) *In the National Interest – Towards a sustainable tourism industry in Malta*. Progress Press, Valletta, Malta.

Ruggieri, G. (2016) Tourism development: concepts and issues. *Tourism Planning and Development* 13(2), 1–2. DOI: 10.1080/21568316.2015.1106776.

Saarinen, J., Rogerson, C.M. and Hall, C.M. (2017) Geographies of tourism development and planning. *Tourism Geographies* 19(3), 307–317. DOI: 10.1080/14616688.2017.1307442.

Schubert, S.F. and Brida, J.G. (2008) Dynamic effects of subsidising the tourism sector. *Tourism Economics* 14(1), 57–80. DOI: 10.5367/000000008783554767.

Sharpley, R. (2020) Tourism, sustainable development and the theoretical divide: 20 years on. *Journal of Sustainable Tourism* 28(11), 1932–1946. DOI: 10.1080/09669582.2020.1779732.

UNDP and WTO (United Nations Development Programme and World Tourism Organization) (1989) *The Maltese Islands Tourism Development Plan – Report prepared for the Government of Malta by the World Tourism Organisation acting as the Executing Agency for the United Nations Development Programme*. WTO, Madrid, Spain.

Walton, J.K. (2009) Histories of tourism. In: Jamal, T. and Robinson, M. (eds) *The SAGE Handbook of Tourism Studies*. Sage Publications, London, UK, pp. 115–129.

Weaver, D.B. (2017) Core–periphery relationships and the sustainability paradox of small island tourism. *Tourism Recreation Research* 42(1), 11–21. DOI: 10.1080/02508281.2016.1228559.

World Economic Forum (2019) *The Travel and Tourism Competitiveness Report 2019: Travel and Tourism at a Tipping Point*. World Economic Forum, Geneva, Switzerland.

# 4 Industrial Tourism in Seville, Spain

**Ana M. Domínguez-Quintero* and Sonia Ancio Alcón**
*University of Seville, Seville, Spain*

## 4.1 Introduction

Industrial tourism is a type of tourism which involves visits to operational companies and industrial heritage (Otgaar *et al.*, 2016). Although it is not a new type of cultural tourism, in recent years it has indeed experienced increasing importance as part of the cultural offering presented by a growing number of destinations (Vargas-Sánchez, 2015). Industrial tourism has arisen thanks to the growing interest of visitors in becoming familiarized with and understanding the recent history of firms in the area, learning about the current and past manufacturing processes of both artisan products and the most advanced technologies, and the need to connect and interact with the population, learn about their customs and traditions and, ultimately, the desire to live real experiences rather than just viewing the scenario (Vargas-Sánchez *et al.*, 2009).

Industrial tourism could be placed within creative tourism, which is one that interacts with the local population, makes no difference between consumption and production, enjoys its own and exclusive experiences of the destination visited, enables tourists to participate therein, and is considered sustainable tourism (Gallego Valiña, 2011; Otgaar *et al.*, 2016). With this type of tourism, an activity is created in which the visitor connects with the culture and nature of the city, coinciding its maximum expression with the fundamentals of sustainable tourism which takes into account the past, current and future economic, social, and environmental repercussions to meet the needs of the visitor (Vargas-Sánchez *et al.*, 2014).

Industrial tourism enables tourists to engage in the evolution of production systems. In its heritage or historical aspect, it offers knowledge and understanding of history and allows visitors to become aware of the economic, social and cultural changes that have occurred. In its aspect of active or operational industry, industrial tourism offers the possibility of visiting companies that are currently active and that may belong to different productive sectors, understood in their broadest sense, of both tangible goods and intangible goods or services (Prat Forga, 2013). Industrial tourism is present in all productive sectors: mainly in the food industry, where it is most widespread; in energy, where companies related to obtaining renewable energy are gaining importance by offering visits to production centres where research is key; in raising livestock; and in crafts and industrial heritage, through touring the heritage that the business fabric has handed down over the years (Frew, 2000; Otgaar *et al.*, 2010; Vargas-Sánchez *et al.*, 2014).

Several studies have pointed out the benefits that industrial tourism can bring both for the company that carries it out, and for the community and the environment where the company

*Corresponding author: anadq@us.es

 *Tourism Planning and Development in Western Europe* (eds K. Andriotis *et al.*)
DOI: 10.1079/9781800620797.0004

resides (Vargas-Sánchez, 2009; Jia, 2010; Prat Forga, 2013). We therefore believe that the study of this issue and the conclusions that can be obtained can be of importance not only for the tourism sector but also for the productive sector and for the community where it can be carried out. However, as Vargas-Sánchez *et al.* (2014) and Vargas-Sánchez (2015) point out, this field of research remains relatively unexplored and there is a clear deficit in the study thereof. The objective of this chapter is threefold. First, to advance deeper into the conceptual knowledge regarding what industrial tourism actually is and what is understood by this type of cultural tourism. The second is to analyse the perception that operational companies in the province of Seville (Spain) that carry out this activity hold of the benefits and costs that this entails. The third objective is to examine, through a case study, the benefits that visits to historical heritage can bring for the economic and social environment of the destination in which this takes place.

The chapter has been structured into eight key points. The first two analyse the concept of industrial tourism and its historical background. The third studies the benefits and costs that starting up this activity brings for companies. The fourth point sets out the context of the study and briefly explains the current situation of industrial tourism in Seville. The fifth point explains the methodology employed in our study. The sixth and seventh points collect the results and the conclusions obtained. Finally, the eighth and last point explains the limitations of the study and future lines of research.

## 4.2 Defining Industrial Tourism: Typology

A review of the literature underlines the lack of unanimity in what is understood by the term *industrial tourism* (Otgaar *et al.*, 2016). Although most authors agree to classify it as a category within cultural tourism, the definitions and terms they use to refer to this type of tourism are various: 'industrial tourism', 'industrial heritage tourism', 'visits to factories', 'factory tourism' or 'industrial archaeology'. The first two of these definitions remain the most commonly used.

According to Edwards and Llurdes i Coit (1996), *industrial heritage tourism* is related to visits to industrial historical heritage which is considered to be disused, abandoned or rebuilt manufacturing facilities, used for tourism purposes. In this same direction, Davis and Rudd (1998) consider that industrial heritage tourism is a subsection of cultural tourism and is associated with the recovery and enhancement of heritage inherited from previous generations.

In Spain, Valls *et al.* (2004) refer to urban *industrial tourism* as that based on the industrial archaeology characteristics of the Industrial Revolution of the late eighteenth and nineteenth centuries. All previous definitions explicitly include visits to industrial historical heritage and therefore exclude visits to operational firms. This conception is the most widely used in countries such as France or Germany, where people associate industrial tourism with industrial heritage, and not with visiting operational firms (Otgaar *et al.*, 2010). In these countries, the terms 'company visits' or 'factory tourism' are often used to refer to visits to active companies.

Anglo-Saxon literature distinguishes between two terms: *industrial tourism* to refer to visits to active companies, and *industrial heritage tourism* to refer to visits to historical heritage.

On the other hand, for Frew (2000), *industrial tourism* involves visits by tourists to operational sites where the core activity of the site is non-tourism oriented. This definition therefore excludes visits to historical heritage and to companies that are not operating. The most recent definitions bring the two perspectives together. For the World Tourism Organization (European Travel Commission, 2005), industrial tourism is a category within cultural tourism whose main characteristic is the visitor's interest in becoming familiarized with and visiting companies, their facilities, manufacturing processes, thematic museums and/or interpretation centres. As stated by Otgaar *et al.* (2010), Jean René Morice presented an interesting definition of industrial tourism at the First European Conference on Industrial Tourism, held in Angers in 2006. According to this definition, industrial tourism includes: (a) visits to companies operating in all economic sectors; (b) visits to industrial heritage as well as to museums where visitors can learn about the previous economic activity; and (c) scientific tourism, which would include visits to museums, cultural centres and scientific centres, where visitors can learn about science.

In our opinion, the definition proposed by Otgaar *et al.* (2010, p. 24) is the most accurate and is therefore that used in this chapter for further analysis. According to this definition, 'Industrial tourism involves visits that enable visitors to learn about economic activities in the past, the present, and the future'.

In analysing this definition further, certain relevant aspects require clarification. The first aspect refers to what is understood by the term *industrial tourists*. In addition to leisure tourists in general, various groups could visit industrial attractions, such as students, journalists, researchers, residents of the same city, associations and professionals. According to Soyez (1993), Otgaar *et al.* (2010) highlight that the key aspect to be considered in industrial tourism is that the visitor's motivation must be an interest in learning about economic activities, and that visitors are not remunerated by the companies they visit. In this respect, it would not be an essential requirement that the visitor travel from another city or from another geographical area to be considered an industrial tourist. People who reside in the same city and even families of the employees can be considered industrial tourists if their interest is to learn about economic activities, since they are moving from their daily environment.

A second outstanding aspect of the definition proposed by Otgaar *et al.* (2010) is that the economic activity that arouses the interest of visitors may have occurred in the past or may occur in the present or in the future. This definition is therefore very broad, and it can include both visits to places of historical heritage and visits to operational firms. In the former, visitors learn ancient production methods, and everything related to the industry, as well as its evolution over time. In the latter type of firm, visitors learn about current production methods and about the research that may be taking place that can guide their evolution in the future. This definition would therefore include what in the Anglo-Saxon literature is called *industrial heritage tourism* and *industrial tourism* or *visits to active firms*. A third aspect that Otgaar *et al.* (2010) emphasize in their definition is that the industry does not have to be defined only as a manufacturing industry. The firms receiving the visits can produce physical products or they can generate services and may belong to any of a wide variety of economic sectors, such as agriculture, winemaking, beer and energy. It is also highlighted that the activity may be specifically oriented towards tourism, since the services provided by these companies can also generate interest among visitors. Therefore, this type of firm includes zoos, interpretation centres and museums. Finally, it should be underlined that companies may be publicly or privately owned. Those that are dedicated to the enhancement and recognition of historical heritage are normally publicly owned.

According to the literature review, it can be affirmed that the most recent definitions distinguish between two types of industrial tourism (European Travel Commission, 2005; Otgaar *et al.*, 2010; Pardo, 2010):

1. *Industrial tourism* or *tourism of operational firms*, which includes visits to active companies, in which their main activity is not tourism and where the main motivation of the visitor is to become familiar with the production processes, the peculiarities of the company, its products and its philosophy.
2. *Heritage industrial tourism*, where the objective of visits usually involves the heritage of old production factories, industrial routes, museums and/or interpretation centres and in this the visitor's motivation is to become acquainted with everything related to the production methods of the past and their evolution up to the present day.

## 4.3 Historical Background

It is not easy to date the beginning of industrial tourism, although De Cauter (1993) does argue that the holding of world fairs in the late 19th century represented an excellent opportunity for visitors to appreciate the importance of industrial production and consumption and, in his opinion, it could signify the origin of industrial tourism. Frew (2000) compiles several examples of visits to production centres in the late 19th century, wherein she cites visits to flower markets and cheese factories in Holland, visits to wineries and chocolate factories in France, and guided visits in the United States to Jack Daniel's distilleries in Tennessee in 1866, and visits to the chocolate factories in Pennsylvania in 1904.

More recently, in the United Kingdom, industrial tourism has been related from its beginnings to the recovery and enhancement of the industrial legacy inherited from the Industrial Revolution and in disuse, and in many cases abandoned, since the end of the Second World War (Otgaar *et al.*, 2016).

Two historic events in the 1960s in the United Kingdom were the prelude to the beginning and development of industrial tourism in that country. The first of these occurred in 1962. The demolition by British Railways of the Doric portico at Euston railway station in London sparked numerous protests from historians and teachers. These protests were the beginning of a movement for the defence of historical heritage. The second major event happened in 1967 when the Ironbridge Valley Museum was created, considered one of the first examples of industrial tourism. This museum, located in the valley of the River Severn, receives approximately 300,000 visits annually, and has served as a revitalizing element and promoter for the development of the entire area. Davis and Rudd (1998) refer to this type of visit as framed within industrial heritage tourism, whereby this type of tourism is a category within heritage tourism.

In Europe, starting in the 1960s, many industrial regions created during the Industrial Revolution, and which at the time were of major importance for the economic development of the area, began to enter a recession. This process was especially intense in the textile and mining industries, whose facilities have subsequently served as a resource for industrial tourism (Hospers, 2002). In Spain, traditionally industrial regions such as Catalonia and the Basque Country suffered this same process and it was precisely in these two communities that industrial tourism emerged in Spain (Cánoves and Prat, 2012). After the deep economic crisis of the 1970s, there was a generalized closure of factories, but at the same time there was an interest in visits to industries, mainly those that were abandoned within the urban space and which were given a tourist pertinence, many of which later became a cultural asset of the city. In this type of tourism, the cities of Bilbao and Barcelona have found a chance to revalue their industrial heritage (Pardo, 2010).

The concept of industrial tourism was extended to include visits to active firms. In 1988 the British Tourist Authority started the campaign 'See Industry at Work' (Otgaar *et al.*, 2010), which is considered to be one of the first initiatives to promote visits to operational firms. This campaign inspired others that, with the same objectives, were launched in other countries such as the Netherlands, Japan, the United States, Italy, France and Australia. In Spain, the National Industrial Heritage Plan was approved in 2011, which is promoting this activity and has become a planning instrument for the future. On the other hand, in October 2013, the Spanish Industrial Tourism Network (RETI) was presented with the objective of recognizing, valuing and promoting industrial tourism. The aim is to market quality industrial tourism and promote a prestigious image of Spanish industry thanks to the related experiences of its visitors. This network groups together the offer of visits to operational firms and industrial heritage.

## 4.4 Positive and Negative Repercussions of Industrial Tourism

The practice of industrial tourism can have positive and negative consequences both for the destination where it is carried out and for the firms that directly exercise it in their facilities. For destinations, industrial tourism is a wonderful instrument to enhance archaeological and historical industrial heritage (Vargas-Sánchez, 2015). It is a real opportunity to generate tourist activity in towns and areas that are traditionally not touristy, but it also serves to provide a quality complementary offer and to relieve the pressure on already established tourist destinations. The attraction of new tourists to the destination or the lengthening of their stay has positive effects on the economy of said destination, by generating higher spending by tourists and therefore higher incomes at the destination.

The establishment of visits in a company can also benefit the progress of other nearby companies to which services can be subcontracted or their demand may be increased by the greater number of tourists. Moreover, visits may also provide a channel that enables collaborations and synergies between the current productive fabric and the tourism industry to be generated, which presents a useful instrument to improve the image of the destination and to position the brand of the destination in the tourism market (Frew, 2000; Otgaar *et al.*, 2016). The negative consequences for destinations associated with industrial tourism tend to coincide with those of tourism in general (overcrowding, noise, litter, etc.) but to a lesser extent than general tourism since industrial tourism is non-mass tourism.

On the one hand, for the firms that receive visits, industrial tourism can provide numerous benefits and drawbacks (Frew, 2000; Vargas-Sánchez, 2009; Otgaar *et al.*, 2016). Among the benefits, the following deserve mention.

1. Industrial tourism is an effective marketing tool for their products. Visitors can create a special bond with the products and with the company, and this special connection, in addition to building customer loyalty, can serve as a means of dissemination through their recommendations to acquaintances and relatives.
2. Visits can allow a closer relationship with customers and a better understanding of their tastes and needs.
3. Firms can obtain extra income generated from visits by selling tickets, selling products and souvenirs and even selling products that have some damage to their packaging.
4. Visits can make firm workers feel more motivated in their jobs, partly because they see that their jobs are recognized by visitors and partly because of the pride of belonging to the company. This can cause workers to increase their productivity.
5. The recognition of the firm by its customers and by the neighbourhood and the pride or satisfaction of its workers can serve as a spur for more people to show interest in working for said company. Visits can therefore generate interest for potential employees.

Regarding the inconveniences and negative repercussions that companies experience in initiating visits to their facilities, the following can be cited.

1. There are costs associated with the initiation of the activity, including the hiring of guides and interpreters or the adaptation of the facilities in accordance with legal and security regulations, such as access, reception area, services, areas specifically designated for the visitors within the production area, car parks and exhibition of the products.
2. Another reason that can prevent or limit visits to factories may be that they make it difficult for the company to function normally, due to the special characteristics of the production process. This can occur in companies where security is especially important and where visits can be dangerous, for example laboratories, energy-producing companies and petrochemical companies. Extreme security measures can constitute a limitation for visits.
3. Another cost associated with establishing visits in a company can be the potential risk of industrial espionage. Treece (1995) (in Frew, 2000) suggests that one of the reasons why the Kellogg's company closed its plants to visitors was precisely to prevent competitors from taking photos of the production process.

Although all these costs must be undertaken by the firms at the opening of the visits, they should be compared with the aforementioned medium- and long-term benefits. Obviously, medium-sized and large firms enjoy more possibilities for this than small companies, and, in this respect, the public administration could play an important role by providing direct legal advice to firms and subsidizing or facilitating access to financing for these initial costs.

## 4.5 Context of the Study: Situation of Industrial Tourism in Seville

First of all, it should be stated that the province of Seville has been chosen since it is where the two authors live and because it also constitutes a major tourist destination within Spain and Andalusia (Prodetur, 2016; Junta de Andalucia, 2020). The province of Seville has an area of just over 14,000 km and a population of 1,957,197 inhabitants at the beginning of the year 2020, making it the largest and most populated province in Andalusia. Its location in the centre of the Guadalquivir valley makes it an area that is particularly well linked both with the other Andalusian provinces and with the remaining Spanish provinces. It has a modern communications network made up of its international airport, a modern high-speed rail network, a network of motorways and a river port that connects it to the Atlantic. In 2019, Seville received 4,292,356 visitors, which represents 13.2% of the total number of tourists who visited the Andalusian region, with an average stay of 3.3 days and an average daily expenditure of €73 (Junta de Andalucia, 2020).

In relation to industrial tourism, the creation in 2014 by the Provincial Council of 'Industrial Tourism of Seville and Province' business meetings should be highlighted. The objective is the exchange of experiences, ideas and initiatives and the promotion of the existing offer in the province of Seville. These meetings are associated with the bestowing of an annual award within the field of industrial tourism in two aspects: visits to operational firms and visits to industrial heritage. That same year, the Seville Provincial Council, through its Prodetur agency, created a catalogue of industrial tourism resources that can be visited in the province of Seville and are available on its website, and which currently shows the data of 149 firms spread over 57 municipalities (Prodetur, 2016). The offer of industrial tourism in the province of Seville is classified into five types of visit, those belonging to the agri-food sector, the industrial and energy sector, the livestock sector, artisans, and firms with industrial heritage. Among the establishments that receive visits to their facilities are some dedicated to the artisan production of oils, wines and spirits, leather derivatives, ceramics, food products, solar-power generation plants and a powerful aeronautical industry.

## 4.6 Methodology

Given that there are two types of industrial tourism, namely visits to active companies versus visits to historical heritage, the use of different methodologies for each type of company has been considered. For the first group of companies, a descriptive analysis was carried out on the perception that operational firms hold regarding the benefits and costs associated with visits and the intention of these companies to maintain such visits. For the second group, the researchers selected the specific case of the firm that was awarded the modality of industrial heritage tourism by the *Diputación de Sevilla* in 2019: the Archaeological Interpretation Centre of Lagar de Osset. This constitutes a fairly representative example of industrial heritage. In this case, an analysis is carried out on the history of the company, its beginnings as an industrial tourism centre, and the repercussions that the visits have had on the nearby economic and social environment. To this end, the information available on the company's website has been employed, together with an in-depth interview with the manager.

Focusing on the first group of companies, the population under study comprises the 127 operating companies that are included in the catalogue drawn up by the Prodetur agency that receive visits at their facilities in the province of Seville. The data were collected during the months of September and October 2020, which coincided with the restrictions on visits to companies due to the Covid-19 pandemic. The questionnaire was sent by email three times, at ten-day intervals, to each of the 127 companies in PDF format so that they could return it completed through a link to a data collection application. Despite the fact that the researchers called each and every one of the companies several times to ask them to please fill out the questionnaire, only 38 of them did so, which represents 29.92% of the population. Based on the literature review (Frew, 2000; Otgaar *et al.*, 2010, 2016), the authors have designed a questionnaire that has two parts. In the first part, the evaluations that firms provide on the positive and negative effects of receiving visits in their company are collected through 16 items that are measured using a 7-point Likert scale, where the value 1 indicates total disagreement with the statement and value 7 indicates total agreement with the statement. Seven specific elements and a global assessment element have been included to assess the perception of benefits:

1. Visits have contributed towards attracting new customers and building loyalty to those they already had.
2. Visits have made it possible to better understand the needs and tastes of customers.
3. Visits have contributed towards improving the image of the company.
4. Visits have facilitated the marketing of the products.
5. Visits have led to an increase in sales.
6. Visits have increased the motivation and productivity of employees.
7. Visits have favoured the activity of nearby companies.
8. Visits in general have improved the performance of the company.

In order to assess the perception of the costs, three specific items and one global assessment item have been used:

1. Visits have hindered the normal functioning of my company.
2. Visits have incurred a high cost for my company.
3. Visits have implied a risk that others will copy ideas from my production process.
4. The costs and difficulties of initiating the visits have been worth it.

Finally, an item has been included regarding their perception of the help of the public administration in the implementation of the visits, and an item for the assessment of their intention to continue with the visits in their facilities. The second part of the questionnaire collects the data related to the company in question. These include ten items: name of the company, municipality in which it is located, sector of activity, founding year of the company, year of initiating visits, number of employees, type of visitors they receive, whether or not they charge an admission fee, whether they carry out tastings and whether they offer their products for sale during the visit. The appendix to this chapter (p. 64) includes the list of companies that participated in this study, the municipality in which they are located and the activity sector to which they belong. For the statistical analysis of the data, the researchers used the SmartPLS v.3.3.3 program.

## 4.7 Analysis of the Results

### 4.7.1 Analysis of the results related to operational companies

In accordance with Hair *et al.* (2018), the measurement model is evaluated by assessing the reliability and validity of composite Mode A constructs of positive and negative effects of a firm receiving visits. Table 4.1 shows that the indicators of both variables meet reliability requirements since, in general, loadings are higher than 0.7. In addition, there are certain items with a loading of 0.4–0.7. These items should be considered for removal if they increase the composite reliability (CR) and average variance extracted (AVE) to a level above the threshold. However, the decision is to retain them to support the content validity of the scale. Table 4.1 shows that the CR is greater than 0.7 and the AVE of the constructs is higher than 0.5, and therefore convergent validity is acceptable (Hair *et al.*, 2018). The fact that CR and AVE surpass the 0.5 threshold confirms that the removal of indicators in the two groups with loadings of 0.4–0.7 was unnecessary.

Table 4.2 shows that both variables achieve discriminant validity following both the Fornell–Larcker and the HTMT90 criteria (Henseler *et al.*, 2015).

The results related to the second part of the questionnaire are summarized in Table 4.3. From these results, it can be observed that 68.42% of the companies belong to the agri-food sector, and that companies that belong to the artisan sector comprise the second group, which represents 15.79% of the sample. Regarding the founding year of the company, it should be noted that 50% of the companies were created after the year 2000 and almost 24% in the last ten years. Regarding the year in which the visits to their facilities began, 50% of the companies opened to visitors as of 2011, a recent year. Regarding the number of employees, it can be observed that 57.89% are micro-companies with fewer than 10 employees, almost 37% are small- and medium-sized companies with between 10 and 249 employees, and only 5.27% are large companies with more than 249 employees. It is therefore remarkable that the majority are micro-enterprises. Regarding the type of visitor,

**Table 4.1.** Measurement model for Mode A composites: weights, loadings, construct reliability and convergent validity.

| Constructs/items | Weights | Loadings | CR | AVE |
|---|---|---|---|---|
| Positive effects | | | **0.908** | **0.591** |
| PE1: visits have contributed towards attracting new customers and building loyalty to those they already had | 0.224*** | 0.826 | | |
| PE2: visits have made it possible to better understand the needs and tastes of customers | 0.175 n.s. | 0.641 | | |
| PE3: visits have contributed towards improving the image of the company | 0.146 n.s. | 0.627 | | |
| PE4: visits have facilitated the marketing of the products | 0.242*** | 0.882 | | |
| PE5: visits have led to an increase in sales | 0.149*** | 0.908 | | |
| PE6: visits have increased the motivation and productivity of employees | 0.165** | 0.597 | | |
| PE7: visits have favoured the activity of nearby companies | 0.196*** | 0.832 | | |
| Negative effects | | | 0.746 | 0.512 |
| NE1: visits have hindered the normal functioning of my company | 0.414 n.s. | 0.669 | | |
| NE2: visits have incurred a high cost for my company | 0.710 ** | 0.927 | | |
| NE3: visits have implied a risk that others will copy ideas from my production process | 0.137 n.s. | 0.479 | | |

Note: *$p$<0.05; **$p$<0.01; ***$p$<0.001; n.s. not significant

**Table 4.2.** Measurement model for Mode A composites. Discriminant validity.

| Fornell–Larcker criterion | | | Heterotrait–monotrait ratio (HTMT) | | |
|---|---|---|---|---|---|
| | PE | NE | | PE | NE |
| PE | **0.769** | | PE | 0.301 | |
| NE | 0.142 | **0.715** | NE | 0.407 | 445 |

Note: the square roots of AVEs are shown in bold.

we must emphasize that most of the companies receive private tourists, organized groups and families. Only 6.58% receive visits from schoolchildren. Regarding whether they charge a fee for the visit, only 40% of companies do so, while 68.42% of the sample carry out tastings, and 73.68% offer their products for sale during the visit.

The basic descriptive statistics for each of the items related to the first part of the questionnaire are listed in Table 4.4. Regarding the assessment that firms make of the positive effects of visits, the most valued effect is the improvement of the firm's image, followed by an improvement in the acquisition and loyalty of customers, a better understanding of the tastes and needs of customers and an improvement in the activity of nearby companies. Regarding their assessment of the negative effects, in general these three items take low values, and always remain lower than the assessments of the positive aspects. Concerning the comparative assessment of whether the costs and difficulties involved have been worthwhile, the majority of the companies answered positively and with a high average value. As for the support that firms have received from public administration, the majority answered that this aid has been very

**Table 4.3.** Main characteristics of the sample.

| Sector of activity | (%) | No. of employees | (%) |
|---|---|---|---|
| Agri-food | 68.42 | Fewer than 10 | 57.89 |
| Artisans | 15.79 | 10–49 | 18.42 |
| Livestock | 5.27 | 50–249 | 18.42 |
| Industrial and energy | 10.52 | 250 and over | 5.27 |
| **Founding year of the company** | **(%)** | **Visitor types** | **(%)** |
| 1682–1950 | 10.52 | Private tourists | 84.21 |
| 1951–2000 | 39.48 | Organized groups | 73.68 |
| 2001–2010 | 26.32 | Families | 71.05 |
| 2011–2020 | 23.68 | Schoolchildren | 6.58 |
| **Year of initiation of visits** | **(%)** | **Entrance fee** | **(%)** |
| 1985–2000 | 13.16 | Yes | 39.47 |
| 2001–2010 | 36.84 | No | 60.53 |
| 2011–2020 | 50.00 | | |
| **Tastings on visits** | **(%)** | **Offer of products for sale** | **(%)** |
| Yes | 68.42 | Yes | 73.68 |
| No | 31.58 | No | 26.32 |

low. Finally, it should be highlighted that most firms have a very strong intention to continue encouraging visits to their establishments.

### 4.7.2 Analysis of the results related to the case study

As noted above, the Archaeological Interpretation Centre of Lagar de Osset received the sixth Seville Province Industrial Tourism Award in the industrial heritage modality, as organized by the *Diputación de Sevilla* in 2019. Four firms opted for this modality, related to those material and immaterial vestiges of organizations or economic activities of the municipalities that, being part of the past and having a clear tourist attraction, are no longer active. The jury assesses the age, tradition, and degree of consolidation achieved in the provision of the industrial tourism service, and the complementary services offered to the visitor (tastings, shop, parking area, etc.). They also assess values that are enhanced by activity, which include the link with the citizens of the environment, the consideration of accessibility and sustainability criteria in the design and provision of the service, the incorporation of innovative elements, the degree of diversification of the offer and excellence in management.

The Archaeological Interpretation Centre of Lagar de Osset in San Juan de Aznalfarache (Seville) opened to visits in 2011. In this centre, archaeological remains from the Turdetan period, such as Roman and Almohad artefacts, can be visited. However, of special interest and importance is that which they have called 'Lagar de Osset'. The archaeological remains have a chronology of the 3rd century BC. At that time, San Juan was a city whose economy was based on the production and export of wine to the entire Mediterranean. Thanks to archaeological research, we can visit the remains of this winery and learn about the operation of a 2300-year-old wine industry. The documented ruins enable visitors to take a tour of all the productive phases of winemaking, from the treading of the grapes to the transfer and storage of wine. On the tour part of the visit, highly interesting areas can be observed, such as the treading ponds, in which the treading of the grapes from which the first must was obtained was carried out. The tour continues going through the press, in which there is a screw-operated lever dam, which

**Table 4.4.** Basic descriptive statistics.

| | Min. | Max. | Mode | Median | Mean | SD |
|---|---|---|---|---|---|---|
| Positive effects | | | | | | |
| PE1: visits have contributed to attracting new customers and building loyalty in old customers | 1 | 7 | 7 | 5 | 5.22 | 1.65 |
| PE2: visits have made it possible to better understand the needs and tastes of customers | 1 | 7 | 5 | 5 | 4.51 | 1.76 |
| PE3: visits have contributed to improving the image of the company | 3 | 7 | 7 | 7 | 6.11 | 1.20 |
| PE4: visits have facilitated the marketing of products | 1 | 7 | 4 | 4 | 4.46 | 1.85 |
| PE5: visits have led to an increase in sales | 1 | 7 | 1 | 4 | 3.72 | 2.13 |
| PE6: visits have increased the motivation and productivity of employees | 1 | 7 | 4 | 4 | 3.86 | 1.88 |
| PE7: visits have favoured the activity of nearby companies | 1 | 7 | 5 | 5 | 4.51 | 1.76 |
| Global valuation: visits in general have improved the performance of my company | 1 | 7 | 4 | 4 | 4.47 | 1.86 |
| Negative effects | | | | | | |
| Item 9: visits have hindered the normal functioning of my company | 1 | 7 | 1 | 4 | 3.49 | 1.94 |
| Item 10: visits have incurred a high cost for my company | 1 | 7 | 4 | 3 | 2.83 | 1.46 |
| Item 11: visits have implied a risk that others will copy ideas from my production process | 1 | 7 | 1 | 3 | 3.16 | 1.85 |
| Global valuation: costs and difficulties involved in initiating the visits have been worthwhile | 1 | 7 | 7 | 6 | 5.28 | 1.59 |
| Valuation of public support | | | | | | |
| Item 13: my firm has had support from the public administration (Junta de Andalucía, Provincial Council or city council) to organize the visits | 1 | 7 | 1 | 2 | 2.97 | 2.01 |
| Intention | | | | | | |
| Item 14: my firm intends to continue organizing visits to its facilities | 1 | 7 | 7 | 7 | 6.19 | 1.52 |

multiplied production and is of great importance for the history of San Juan. Finishing in the fermentation and decantation pools or ponds, the must, once trodden and/or pressed, flowed into these. The must remained in these pools for approximately a week, until the fermented wine and the impurities had been removed. Once the must was fermented and decanted, it would be left to age in these same containers until it was time for them to be packed and the required finish was then given for each type of wine. The tour also includes a visit to the winery, in which the wine-ageing process was carried out in underground spaces at a depth of 3.5 metres. Visits are adapted in accordance with the audience profile:

1. Schools, for which workshops are held related to the historical past of the municipality, such as gymkhanas and mosaics.
2. Specialized groups (such as historians and archaeologists), to which information of a more specific nature is provided depending on the group and supported by the attendance of an expert archaeologist.
3. Foreigners, who are offered tours in English. At the end of the visit, the company offers a tasting of wines from the region and visitors can buy these local wines and souvenirs in the shop.

Those responsible for this company depend on the city council, and it should be noted that this interpretation centre has reported the most significant benefits to the municipality:

1. It has served to attract visitors and to make the destination known both in Andalusia and across Spain.
2. It has improved the perceived image of the destination by both residents and visitors.
3. It has improved employment in the area.
4. It has favoured the activity of other nearby companies.
5. In general it has boosted the local economy.

## 4.8 Conclusions

The theoretical review of the literature has allowed us to expand our knowledge of the concept of industrial tourism and to separately clarify its active and heritage aspects. In this way, the first of our three objectives is satisfied. The data collected in our study confirm the enormous surge that industrial tourism has been experiencing in recent years. This can be deduced when verifying that the number of firms that receive visits to their facilities in the province of Seville has doubled in the last ten years. This coincides with previous studies (Frew, 2000; Otgaar *et al.*, 2010, 2016; Vargas-Sánchez, 2015).

Regarding the positive and negative effects perceived by the active companies, we have confirmed the reliability and validity of both constructs through a statistical analysis. To the best of our knowledge, there are no previous practical studies in this regard. Regarding the evaluation of benefits received by active companies, most of them value the benefits associated with visits more positively and with greater intensity than they do with the incurred costs. Among the most valued benefits are the improvement of the image of the company, the acquisition and retention of new customers, a better understanding of the tastes and needs of customers and the improvement in the performance of other nearby companies. The latter shows a positive impact, and the multiplier effect of this type of tourism arises in the local economy through the improvement in the operation of nearby companies. Regarding the valuation of costs, valuation in general is moderately low, which implies that companies do not perceive these costs as an insurmountable obstacle. In fact, most companies state that the initial costs and difficulties in setting up visits have been worthwhile. In accordance with these findings, most of the companies show a strong intention to maintain visits to their company in the future. These results confirm previous theoretical and practical studies (Pardo-Abad, 2004; Vargas-Sánchez, 2009, 2015; Otgaar *et al.*, 2016). We believe that this subsection of our work satisfies our second objective, and has enabled us to ascertain the benefits and costs that industrial tourism activity brings for the companies undertaking such activity in Seville.

Finally, it should be highlighted that firms perceive the help they have received from the public administration as being very poor. Even though industrial tourism is considered a non-mass and sustainable type of tourism, with enormous benefits for industries and for the local destination, policymakers fail to pay sufficient attention to this activity. By bearing in mind that, in Seville, almost 60% of the companies that are open to visits are micro-firms, it can be deduced that they have made an enormous effort to initiate this activity. To the best of our knowledge, there are no previous studies on this issue, and therefore we believe that it would be opportune to research in this direction in greater depth. Support for the initiation and maintenance of this type of tourism could be a significant economic policy instrument for the revitalization of a destination's economy.

On the other hand, the study of the specific case of the Archaeological Interpretation Centre and the interviews with its managers has allowed us to confirm the benefits that this type of interpretive centre can yield for the economic and social environment of the area. Given that most of these centres are publicly managed, it is evident that the recovery of heritage and the promotion of these interpretive centres could present another valid instrument for the revitalization of the economy of a destination. This subsection of our work satisfies our third objective.

### 4.8.1 Research limitations and future lines of research

Our work has several limitations. The main drawback involves the relatively small size of the sample of operating firms, since the special Covid-19 circumstances in which we find ourselves have prohibited access to the data of a greater number of companies. The second limitation is the geographical restriction to the province of Seville. We should like a future study to have a greater amount of data from Seville firms and data from other Andalusian or Spanish provinces. This would enable wider generalizations to be made regarding our conclusions and would also allow the application of other types of statistical or econometric techniques with stricter prior conditions. For example, a larger sample would allow us to estimate, by employing a structural equation model, the influence of the benefits and costs perceived by companies on their intention to continue with visits to their facilities. With this model, it would also be of major interest to analyse the moderating role of public sector support.

On the other hand, there is a major gap in the study of the benefits and costs that industrial tourism in its double aspect or modality can bring for a local economy. Since there are numerous types of statistics available on this specific type of cultural tourism, an effort in this respect would enable an empirical study to be carried out in greater depth at both the micro- and macroeconomic level.

## References

Cánoves, V. and Prat, J.M. (2012) Associations as the backbone of the development of industrial tourism: the XATIC network and industrial tourism in Catalonia. *Tourism Analysis Magazine* 13, 49–56 (in Spanish).

Davis, J.A. and Rudd, M.A. (1998) Industrial heritage tourism at the bingham canyon copper mine. *Journal of Travel Research* 36(3), 85–89.

De Cauter, L. (1993) The panoramic ecstasy: On world exhibitions and the disintegration of experience. *Theory, Culture & Society* 10(4), 1–23.

Edwards, J.A. and Llurdes i Coit, J.C. (1996) Mines and quarries industrial: Heritage tourism. *Annals of Tourism Research* 23(2), 341–363.

European Travel Commission (2005) *City tourism and culture: the European experience*. World Tourism Organization, Madrid, Spain.

Frew, E.A. (2000) *Industrial Tourism: A Conceptual and Empirical Analysis*. Victoria University of Technology Melbourne, Victoria, Australia.

Gallego Valiña, M.A. (2011) Industrial Tourism. *Journal of Leisure and Tourism* 4, 117–138 (in Spanish).

Hair, J.F., Sarstedt, M., Ringle, C.M. and Gudergan, S.P. (2018) *Advanced Issues in Partial Least Squares Structural Equation Modeling*. Thousand Oaks, CA, Sage, London, UK.

Henseler, J., Ringle, C.M. and Sarstedt, M. (2015) A new criterion for assessing discriminant validity in variance-based structural equation modeling. *Journal of the Academy of Marketing Science* 43(1), 115–35. DOI: 10.1007/s11747-014-0403-8.

Hospers, G.-J. (2002) Industrial heritage tourism and regional restructuring in the European Union. *European Planning Studies* 10(3), 397–404. DOI: 10.1080/09654310220121112.

Jia, Y. (2010) Analysis and suggestions on Chinese industrial tourism development. *International Business Research* 3(2), 169–173. DOI: 10.5539/ibr.v3n2p169.

Junta de Andalucia (2020) *Institute of Statistics and Cartography of Andalusia. Statistics* (in Spanish). Available at: www.juntadeandalucia.es/institutodeestadisticaycartografia/temas/index-est.htm (accessed 14 July 2020).

Otgaar, A.H.J., Van der Berg, L., Berger, C. and Feng, R. (2010) Industrial tourism. In: *Industrial Tourism. Where the Public Meets the Private*. Routledge, Taylor and Francis Group, London, UK.

Otgaar, A.H.J., Van der Berg, L., Berger, C. and Feng, R. (2016) Industrial tourism. In: *Industrial Tourism: Opportunities for City and Enterprise*. Routledge, Taylor and Francis Group, London, UK. DOI: 10.4324/9781315588438.

Pardo-Abad, C.J. (2004) The reuse of industrial heritage as a tourist resource. Geographic approach to industrial tourism. *Works of the Catalan Society of Geography* 57, 7–32 (in Spanish).

Pardo, C. (2010) Industrial heritage in Spain: tourism analysis and territorial significance of some recovery projects. *Bulletin of the Association of Spanish Geographers* 53, 239–266 (in Spanish).

Prat Forga, J.M. (2013) Industrial tourism as an element of revaluation of the territory: an analysis from the social relations present in the destinations: Analysis of cases in Catalonia, Scotland, and Alsace. Autonomous University of Barcelona. *TDX (Doctoral Thesis in Xarxa)* (in Spanish). Available at: http://www.tdx.cat/handle/10803/120536 (accessed 10 December 2019).

Prodetur (2016) *Industrial Tourism in the province of Seville*. Provincial Council of Seville (in Spanish). Available at: www.juntadeandalucia.es/turismoydeporte/publicaciones/143543093_01.pdf (accessed 20 September 2019).

Soyez, D. (1993) Cultural tourism in industrial landscapes. In: Becker, C. and Steinecke, A. (eds) *Cultural Tourism in Europe: Growth Without Borders* (in German). European Tourism Institute at the University of Trier, Trier, Germany.

Treece, J.B. (1995) Will plant tours go the way of the tail fin? *Business Week* 10 April.

Valls, J.F., Bustamante, X., Guzmán, F. and Vila, M. (2004) *Management of sustainable tourist destinations*. Gestión 2000, Barcelona, Spain (in Spanish).

Vargas-Sánchez, A. (2009) *Mining industrial tourism in the province of Huelva*. University of Huelva, Resource CD (in Spanish).

Vargas-Sánchez, A. (2015) Industrial heritage and tourism: a review of the literature. In: Waterton, E. and Watson, S. (eds) *The Palgrave Handbook of Contemporary Heritage Research*. pp. 219–233.

Vargas-Sánchez, A., Plaza-Mejía, M. de los Á. and Porras-Bueno, N. (2009) Understanding residents' attitudes toward the development of industrial tourism in a former mining community. *Journal of Travel Research* 47(3), 373–387. DOI: 10.1177/0047287508322783.

Vargas-Sánchez, A., Porras-Bueno, N. and Plaza-Mejías, M.A. (2014) Industrial tourism. In: Jafari, H.X. (ed.) *Encyclopedia of Tourism*. pp. 470–472.

## Appendix: List of participating companies

| Name | Municipality | Activity sector |
|---|---|---|
| Canal Sur Radio y Televisión, SA | San Juan de Aznalfarache | Industry and energy |
| Bodegas Góngora | Villanueva del Ariscal | Agri-food |
| Basilippo Calidad Gourmet SL | El Viso del Alcor | Agri-food |
| Coop. Oleand S. Coop. Andaluza | La Puebla de Cazalla | Agri-food |
| WellDone Lácticos | Espartinas | Agri-food |
| Ingeoliva Cortijo el Puerto | Lora del Río | Agri-food |
| Antigua Fábrica de Mantecados El Gamo | Estepa | Agri-food |
| Destilerías de Constantina | Constantina | Agri-food |
| Manzanilla Olivé | Utrera | Agri-food |
| Hacienda Guzmán | La Rinconada | Agri-food |
| Forja Lebrija | Lebrija | Artisan |
| Destilería la Flor de Utrera SL | Utrera | Industry and energy |
| Instituto Andaluz de Logística, S.L. | Valencina de la Concepción | Agri-food |
| Cervezas Artesanas del Aljarafe, SL | Sanlúcar la Mayor | Agri-food |
| Ruiz Torreño SL | Espartinas | Agri-food |
| Queseria Dehesa Castilblanco | Castilblanco de los arroyos | Agri-food |
| Discover On Turismo Industrial | Sevilla | Industry and energy |
| Inés Rosales SAU | Huevar | Agri-food |
| Paco Parra | Sevilla | Artisan |
| Mosaicos Casariche, SLU | Casariche | Artisan |
| Cobre Las Cruces SAU | Gerena | Industry and energy |
| Escuela de Formación de Artesanos | Gelves | Artisan |
| Yeguada Francisco Olivera | Alcalá de Guadaíra | Livestock |
| Obrador Valle de la Osa | Constantina | Agri-food |
| Bodegas F.Salado S.L. | Umbrete | Agri-food |
| Oleoestepa | Estepa | Agri-food |
| Puerto de Indias | Carmona | Agri-food |
| Craop. Estepa | Estepa | Agri-food |
| Flor de Cazalla | Cazalla de la Sierra | Agri-food |
| Huerta Camporico SL | Burguillos | Agri-food |
| Dehesa la Calera SL | Gerena | Livestock |
| Quesos Los Vázquez, SL | Castilleja del Campo | Agri-food |
| Manuel Acosta SL | Morón de la Frontera | Artisan |
| La Alameña | El Álamo (El Madroño) | Agri-food |
| Carruajes Romero SLU | Lebrija | Artisan |
| Postres y Lacteos Mare Nostrum, SL | Castilblanco de los Arroyos | Agri-food |
| Mantecados El Dulce Nombre | Estepa | Agri-food |
| Destileria Rigo, SC | Fuentes de Andalucía | Agri-food |

# 5 Salt pans: An Indissociable Natural and Cultural Heritage – A Comparative Study Between Aveiro, Portugal and Guérande, France

**Margarida Ferreira da Silva[1,2*], Helena Albuquerque[3,4], Filomena Martins[2,5] and Gildas Buron[6]**

*[1]GOVCOPP, Department of Economics, Management, Industrial Engineering and Tourism, University of Aveiro, Campus Universitário de Santiago, Aveiro, Portugal; [2]CESAM, Department of Environment and Planning, University of Aveiro, Campus Universitário de Santiago, Aveiro, Portugal; [3]REMIT, Department of Tourism, Heritage and Culture, University Portucalense, Porto, Portugal; [4]GOVCOPP, University of Aveiro, Aveiro, Portugal; [5]GOVCOPP, Department of Environment and Planning, University of Aveiro, Campus Universitário de Santiago, Aveiro, Portugal; [6]Musée des Marais Salants, Cap Atlantique, Batz-sur-Mer, France*

## 5.1 Introduction

Natural and cultural heritage are significant assets for regions, as they allow tourists to differentiate destinations and create an image of them, offering a competitive advantage. Also, they are of great value for local communities, promoting a sense of belonging and making such communities proud of their region.

In some coastal areas (e.g. Piran, Slovenia; Trapani, Italy; Guérande, Île de Ré and Île de Noirmoutier, France; and Aveiro, Figueira da Foz, Olhão, Tavira and Castro Marim, Portugal), salt is still produced with artisanal methods in traditional salt pans located inside wetlands and natural parks. The salt pan landscape is defined by Kortekaas (2004, p. 199) as 'saltscapes': 'a type of cultural landscape formed in salt-making areas, combining saline seminatural habitats and cultural values related to salt making activity'. These anthropogenic landscapes are managed by salt workers and play an important role in the preservation of saltscapes, namely concerning biodiversity conservation, with the attribution of different protection status to these areas, such as Special Protection Area (SPA) and Site of Community Importance (SCI), both in the scope of the Natura 2000 network area (EC, 2005) and classified as wetland of international importance, integrating Ramsar sites (Ramsar, 2016).

These areas have special conditions for nesting, feeding and resting of different bird species, being an important replacement habitat. Salt activity, practised for centuries in

*Corresponding author: a.silva@ua.pt


DOI: 10.1079/9781800620797.0005

an artisanal manner, has contributed to the maintenance of these areas. However, from the end of the 1970s, salt pans along the Atlantic European coast were progressively abandoned, which relates to numerous factors, such as competition with low-cost salt (mechanically produced), high maintenance costs of the salt pans, lack of skilled labour and low profitability in product selling (Silva, 2010; Martins *et al.*, 2013; Kortekaas, 2017).

With the abandonment of salt activity and consequent degradation of these areas, there is a risk of losing natural, cultural and historical heritage that is associated with traditional salt pans and salt production. Currently, traditional salt pans are exposed to natural and human threats: environmental threats such as coastal erosion, pollution and flooding, and human threats such as vandalism, commodification and tourism pressures (Martins *et al.*, 2020). Some authors specifically emphasize the need to control changes in coastal areas that are threatening geographical and cultural aspects and that impact local communities (Chylińska, 2019). Regarding tourism, if it is not sustainably managed, there could be damage to salt pans, which are sensitive biotypes. Also, tourism could lead to negative effects on cultural heritage associated with salt production (Rodrigues *et al.*, 2011).

Additionally, other authors have stated that tourists search for cultural products valued by authenticity (Domínguez-Quintero *et al.*, 2020; Pafi *et al.*, 2020) and by originality (Xie *et al.*, 2020). Kim *et al.* (2019) emphasized the relevance of intangible heritage in cultural authenticity – for example, to differentiate a tourist destination or to promote heritage maintenance in a community. Although there is no consensus regarding the concept of authenticity among researchers, some authors point out that the satisfaction when experiencing heritage is not dependent on literal authenticity, but on *perceived* authenticity (Chhabra *et al.*, 2003).

Traditional salt pans, designated as saltscapes, are anthropogenic cultural landscapes built several centuries ago to produce artisanal salt, which is an activity with an important tangible (e.g. wood tools, baskets, warehouses) and intangible (e.g. salt-making know-how) cultural heritage. Salt production allows visitors to experience the benefits of this centuries-old natural and cultural heritage. Moreover, it contributes to the preservation of natural and cultural heritage related to salt-making activity, either tangible or intangible, since without salt-making activity, the landscape could either disappear or would need to be artificially maintained (Kortekaas, 2017).

With that in mind, this natural and cultural heritage is the image of destinations such as Guérande (France) and Aveiro (Portugal), although each case has different lines of action. Guérande has focused its efforts on the preservation and dynamization of artisanal salt production, having tourism as a complementary activity. Guérande was chosen as a case study for its success in preserving a secular activity and its natural and cultural heritage, dignifying a profession and its community. Also, this approach has allowed a balance to be found in the integration of complementary activities, through a participative strategy with a diversified group of stakeholders. On the other hand, the central point in the case of Aveiro is the development of tourism activities in salt pans, rather than prioritizing salt-making activity, with the risk of it being only represented in a 'museum'. Therefore, the Guérande approach is a good model for the Aveiro salt pans to follow, as Aveiro has had some difficulty in preserving salt-making, and in defining an integrated and participatory strategy. Additionally, the Guérande model is useful for other sites still producing salt in an artisanal manner, contributing to salt pan and salt-making heritage conservation.

One of the questions discussed within this study is whether traditional salt pans could subsist if driven by individual initiatives. The chapter aims to understand whether these initiatives could exclusively focus on tourism activity, even if they do not have an integrative strategy, involving a set of stakeholders, for preservation and development of salt-making in the context of sustainable development. Thus, the main objectives of this chapter are first to identify and explore the main best practices of the Guérande salt pans (i.e. salt-making as a main activity with complementary activities) and second to compare Guérande with Aveiro, in order to evaluate how these best practices could serve as guidelines.

The chapter is divided into several parts. It presents a theoretical approach regarding the importance of salt in history, the concept of sustainable tourism and the development of sustainable tourist activities in salt pans, followed by the methodology and the case studies. Lastly, it provides guidelines so that policymakers and other stakeholders can reflect on the potential of salt pan areas to create sustainable multifunctional synergies.

## 5.2 Literature Review

### 5.2.1 Salt and traditional salt pans as cultural and natural heritage

Over the centuries, salt has played an important role in history, with repercussions in the global economy and the culture of people, promoting trade between different countries. The Phoenicians used salt to conserve fish (salt fish), the Egyptians to mummify bodies and the Romans paid their soldiers with salt, giving rise to the word 'salary'. In classical antiquity, salt was described both in its uses in everyday life and in its religious aspect (Dias, 2005). The Bible has more than 30 references to salt, such as being used for purification or associated with the baptism rite (Dias, 2005). Also, there are several references to salt in the history, symbolism, etymology and toponymy associated with the organization of events and tourist activities (Amorim, 2001; Dias, 2005; Kortekaas, 2017).

Salt was the development device for many European cities like Venice in Italy, Merkers in Germany, Tuzla in Bosnia-Herzegovina, Salzburg in Austria and Wieliczka in Poland (Hocquet, 1982; Ritz, 1996). In a more controlled way, activities related to the production of salt also played an important role in various regions of the Iberian Peninsula, coastal areas of France, the Adriatic Sea and the Balkans, not only as a vehicle for economic growth but also as an engine of the development process (Silva, 2010).

Salt also played an important role in the Portuguese economy. The oldest documents known with reference to Portuguese salt are from the 10th century. Bastos (2009) notes that in AD 929 some salt pans were sold to the abbot of a monastery in Ovar, while Amorim (2001) and Rodrigues *et al.* (2011) mention a donation of land and salt pans in Aveiro, made by the Countess Mumadona, to the monastery of San Salvador, in Guimarães, in AD 959. At that time, nobles, monasteries and certain corporations owned most of the salt pans (Amorim, 2001).

Despite its importance in history, local production of salt begun to decline between the 1950s and the 1970s. Several reasons could explain this, namely the popularization of refrigerators in the 1950s and the subsequent urbanization pressures that led to the occupation of salt pans by infrastructures such as highways and roads. Also, salt pans started to be used for aquaculture or for tourist purposes (Sainz-López, 2017; Martins *et al.*, 2020). However, salt is not just the product itself, it is also the salt-making area, which links cultural and natural features, presenting different landscapes both among the salt pans or between a salt pan and its region (Amorim, 2001; Silva, 2010).

Salt can be produced in different locations: at the surface (e.g. inland salinas, coastal salinas and salt lakes) or underground (e.g. open cast mines, rock mines and solution mining), and in different types of landscape, such as mountains, slopes, terraces, valleys, wetlands, etc. Wetlands can be coastal (shore, lagoon, marsh, estuary) or inland (built wetland, saline river, stream, lake, peat) (Kortekaas, 2017).

Traditional salt pans, which are located within coastal wetlands, are considered a cultural landscape integrating human, cultural and natural features that are indissociable and interdependent. These features work together to achieve and maintain salt pans in a sustainable and balanced way (Kortekaas, 2017). According to Kyvelou and Gourgiotis (2019) it is difficult to have a clear differentiation regarding what is a natural or a cultural landscape.

These traditional landscapes represent the identity of a community and 'have the power to tell a story' (Kortekaas, 2017, p. 56). Law normally protects natural heritage and protection measures exist in many European countries. Nevertheless, this is not always applied to cultural heritage (Kortekaas, 2017). Two main questions emerged from the necessity of maintaining traditional salt pans in production: the first relates to it being more difficult to have an artificial hydrological regime than a system operated by salt workers (Korovessis and Lekkas, 2009), and the other relates to intangible heritage preservation (e.g. traditions, work methods, etc.) associated with cultural landscapes (Katelieva *et al.*, 2019).

### 5.2.2 Importance of tourism as a sustainable complementary activity in traditional salt pans

Tourism is an economic activity that relies heavily on the environment and its quality, as well as on heritage (i.e. natural and cultural). In fact, both environmental quality and heritage are major factors that determine the level of attraction and interest of an area. However, in sensitive areas such as salt pans, tourism activity can contribute to environmental degradation. Several authors and the United Nations World Tourism Organization (UNWTO) have been drawing attention to the negative impacts of tourism if this activity is poorly managed (Postma and Schmuecker, 2017; UNWTO, 2018).

Consequently, sustainability as a key issue in tourism has become increasingly important. Several opinions have emerged pointing out that sustainable tourism should not be considered a typology of tourism: on the contrary, some authors argue that most types of tourism should include sustainability as a development goal (Asmelash and Kumar, 2019) or that sustainability could be a policy in itself (Bramwell, 2015; UNWTO, 2021). Bramwell *et al.* (2017, p. 1) point out that, currently, there seems to be 'a normative orientation that seeks to re-direct societal systems and behaviour on a broad and integrated path toward sustainable development'. According to Higgins-Desbiolles (2018), tourism needs to be analysed from a broader 'holistic sustainable development' perspective, and not just from that of a reduced notion of controlling tourism. The author points out that a solid human system should be constructed based on 'eco-socio-cultural diversity', just as resilient ecosystems are based on biodiversity (2018, p. 159).

Concerning the aforementioned analysis, this study identifies tourism as a sustainable complementary activity in salt pans. Effectively, tourism in these areas could be relevant in contributing to several sustainable development goals (SDGs), both directly and indirectly, with the possibility of promoting local culture and products and also of creating new, related jobs (i.e. targets 8.9 and 12.b) (UNWTO, 2017). In addition, several authors consider that sustainable tourism should integrate all stakeholders in a participatory process regarding tourism decision-making to achieve cultural equity and sociocultural tourism outcomes (Bramwell, 2015; UNWTO, 2021).

Globalization of tourism activity creates extra challenges to planning and governance, since it is not limited to public policy, which means power is more diffuse (Boluk *et al.*, 2019). This idea is connected to Hall's (2011a). Hall suggests four current categories of governance:

1. Political hierarchies (state governance).
2. Markets (recognizing the power of private economic actors and their associations).
3. Networks (through public–private partnerships and associations).
4. Communities (governance arising from the local level with direct public involvement) (Boluk *et al.*, 2019, p. 857).

These categories are also valid for salt pans, where local communities should be included in the governance process (Siakwah *et al.*, 2020).

As traditional salt pans are normally in biologically sensitivity coastal wetlands, tourism activities should follow the guidelines proposed by the Secretariat of the Ramsar Convention on Wetlands (Ramsar and UNWTO, 2012). Thus, tourism activities should contribute to natural and cultural heritage conservation, while respecting local communities' sociocultural authenticity. Furthermore, other authors have pointed out the relevance of collaborative strategies to achieve balance between sustainability (environmental, economic, and sociocultural) and the activities developed by different stakeholders (Jamal and Stronza, 2009; Waligo *et al.*, 2012; Carlisle *et al.*, 2013).

## 5.3 Methodology

The methodology used in this study was a comparative case study analysis, the data for which was collected by an analysis of existing policy documents (e.g. *Strategic Vision for the Centre Region 2030*, CCDR-C, 2020; *Polis Litoral Ria de Aveiro – Integrated Requalification and Valorisation of the Coastal Border*), reports (e.g. projects reports) and academic works (masters and PhD theses). Descriptive and exploratory analysis methods were then used for the Guérande and Aveiro salt pans (Yin, 2013).

The case studies were selected from information collected on previous research projects (i.e. Ecosal ATLANTIS and Salt of Atlantic) and from previous academic works (Silva, 2010; Kortekaas, 2017) looking at different models of salt pan development. As noted above, the Guérande salt pan is a well-established area that has focused on salt-making as its main interest, and tourism as a complementary sustainable activity, while Aveiro is represented as an area that has suffered a decline over time, focusing as it does on the exploration of salt pans for tourism purposes in recent years. Nowadays, this area is one of the most important coastal urban regions in the Portugal Centre Region, with an increasing number of visitors.

Although the case study approach is criticized by some authors, it is still considered important (Idowu, 2016), since it provides critical examples from the real world that can exemplify certain representative patterns or issues. Despite the limitation of the method, the literature review and the detailed analysis of both case studies might offer insights on how tourism as a tool for sustainable development and participatory governance can positively affect traditional salt pans (Yin, 2013).

## 5.4 Case Studies

Salt pans and traditional salt extraction have a natural, cultural and historical heritage that allows the development of a set of activities linked to these values, namely touristic, leisure, recreational, scientific, pedagogic, historical, ethnographic, health and wellbeing activities that promote territorial revitalization and enhancement (Martins *et al.*, 2020). Some are supported by museums, private companies or research centres, while others are promoted by local or associated producers, which is the case

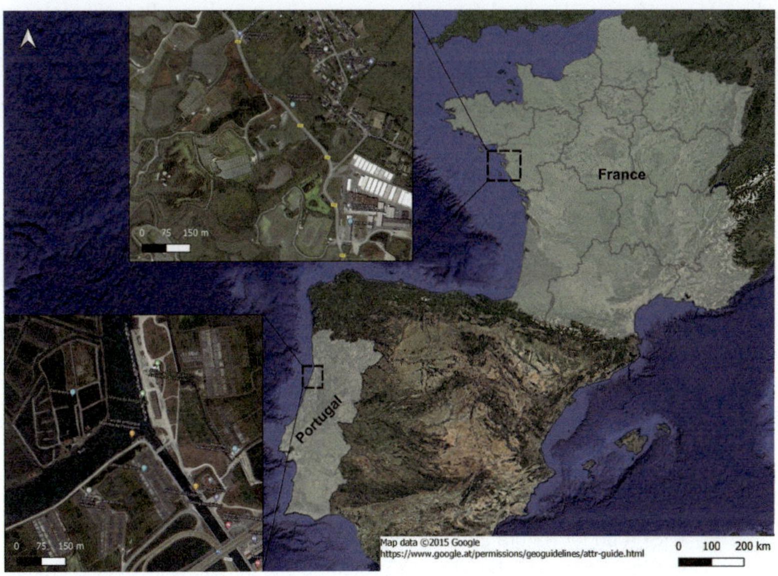


**Fig. 5.1.** Aveiro (Portugal) and Guérande (France), geographical context. (© Google, 2015, www.google.at/permissions/geoguidelines/attr-guide.html.)

with Guérande. Fig. 5.1 shows the geographical context for Aveiro and Guérande.

### 5.4.1 Guérande

The Guérande salt pans are on the Atlantic French coast, in western Loire-Atlantique, and are part of the Agglomeration Community of Guerandaise Peninsula (Communauté d'Agglomération de Presqu'île Guérandaise – CAP). CAP is formed from 15 municipalities, nine of which have salt pans located in Guérande and Baie de Mesquer, integrating Pays de la Loire and south Bretagne (Fig. 5.1). The area is a large wetland occupying 2000 hectares with 1176 hectares of salt marshes, of which 765 hectares consist of salt production (Ménanteau and Nedelec, 2007; Kortekaas, 2017).

Local traditional salt production faced a long period of decline between 1840 and 1960. However, it was between 1950 and 1970 that artisanal salt-making activity faced a high risk of disappearing. Urban development plans, such as roads, emerged, threatening the Guérande salt pans, and a social movement (i.e. academics, salt workers, environmentalists, etc.) acted against these projects and succeeded in preventing their implementation. An organizational structure was created and a professional school for salt workers set up in the late 1970s. This scenario changed in the late 1980s with the development of a strategic marketing study, which focused on three key areas for the revitalization of traditional salt production activity:

1. Classifying the site as a protected area.
2. Dignifying the profession.
3. Increasing product value.

The strategy was carried out by the salt workers' cooperative and included the development of dynamics/partnerships with private and public stakeholders (Table 5.1) (Le Guerandais, 2021).

Specifically, the role of the Societé Coopérative Agricole (SCA) Salines de Guérande has been of key importance. This Guérande salt group, created in 1972, was changed to an agricultural cooperative society in 1988 after the previously mentioned strategic marketing study (1987). The main function of the cooperative, which currently employs 75 people, is to sell the salt produced by its members, who determine the selling price annually. This price-setting guarantees the salt workers a remuneration that allows them to live with dignity (Le Guerandais, 2021).

Alongside sales activity the cooperative has very strong associative relationships, such as with the Terre de Sel visitor centre (see below). The cooperative promotes a spirit of solidarity among its members. For example, every year they help each other in the restoration of salt

**Table 5.1.** Strategic areas for salt pans in Guérande. (Based on Silva, 2010; CAP Atlantique, 2021a; Le Guerandais, 2021).

| Strategic area | Actions |
|---|---|
| Product | • Quality label: Nature & Progrés, Label Rouge<br>• Certification: ISO 9001 and IFS (International Food Standard)<br>• Protected geographical indication (PGI) |
| People | Officially recognized training for salt workers in order to dignify the profession. Following the training, given over two years, salt workers obtain a professional farm management certificate (since 1995 BPREA – Brevet Profissionel Responsable Entreprise Agricole – Activité Salicole) |
| Place | • Zone Naturelle d'Intérêt Écologique, Floristique et Faunistique (natural area of ecological, fauna and flora interest) (ZNIEFF)<br>• Wetlands of international importance under the Ramsar Convention<br>• Important Bird and Biodiversity Area (IBA)<br>• SPA under Natura 2000 Network<br>• Classification of the site as a protected site of natural interest in France<br>• Possibility of being included as a biosphere reserve and on the list for World Heritage (UNESCO) |

pans, which is necessary due to weather and natural conditions (e.g. restoring the dikes). Another common practice is for experienced producers to help set up new producers. In this aspect of solidarity, the cooperative has created the Association Univers Sel, which provides technical support to salt producers around the world, for example in some African countries (Le Guerandais, 2021; Univers-sel, 2021).

Currently, approximately 300 salt workers have their profession recognized and there is a constant concern about their training, a decisive factor for attracting young people. The majority (i.e. 220 salt workers) are members of the cooperative, while the rest are independent workers. The average age of current producers is 46 years old and 75% come from other regions. The cooperative and the Salt workers Union (Syndicat des Paludiers) offer good conditions for new salt workers by providing salt pans and guidance to apply for financial support, since they can be considered 'young farmers'. The activity creates 120 permanent jobs and 300 seasonal jobs. Guérande salt pans produce around 16,000 tonnes of salt and 700 tonnes of *fleur de sel* each year with a revenue of €24 million (Chambre d'Agriculture Loire-Atlantique, 2011; Kortekaas, 2017; CAP Atlantique, 2021b).

In order to consolidate the development strategy, a number of stakeholders (e.g. CAP Atlantique, Chambre d'Agriculture de Loire Atlantique, University of Nantes, APROSELA – Association pour la Promotion du Sel de L'Atlantique, Cerqua – Centre de Développement des Certifications des Qualités Agricoles et Alimentaires, LPO – Ligue pour la Protection des Oiseaux – Loire Atlantique, Sites remarquables de gout and Cuisineries Gourmandes) have contributed in each area of action: planning and policies, research, nature conservation, training, tourism, gastronomy and product quality and strengthening. They have developed a set of relevant actions that have had several positive impacts on the salt pans' activity, grouped into three strategic areas: place, people and product. The cooperative has also established several partnerships over the years regarding quality and promotion of the salt product (Le Guerandais, 2021).

In addition to salt-making, tourism activity has become of key relevance for the dynamics and sustainability of the Guérande salt pans. Tourist demand in the region is associated with the brand Presqu'île Guérandaise. The brand benefits from its proximity to the Bretagne region and its prestige, as well as to the diversified offer in Guérande with a significant natural, cultural and medieval heritage. It also benefits from the maritime heritage of St Nazaire, located 20 kilometres from Guérande, and from the renowned resort of La Baule. Guérande is a city with a population of 16,681 people that receives approximately 1.5 million visitors each year (Ville de Guérande, 2021).

Tourism is responsible for one fifth of generated incomes of the region, employing more than 8,100 people (CAP Atlantique, 2013; 2021c). Initially, salt workers perceived the activity as negatively affecting their work, but currently they see tourism as an educational and a complementary activity, especially due to the contribution of three visitor centres, which received 109,422 visitors in 2019 (Buron, 2021): the Musée des Marais Salants, the Maison des Paludiers and the Terre de Sel (Fig. 5.2). In the past, the three acted as competitors, but they are now complementary in the work and services they offer (they are less than 10 kilometres apart) (Silva, 2010; Kortekaas, 2017). Their tourist offers include a set of activities that can be developed over the year to minimize the seasonal effects of salt-making. However, nonguided visits are avoided in order not to disturb the salt-making activity. Some independent salt workers (i.e. not cooperative members) promote their own guided visits as long as they are scheduled (Le Drogo, 2006; CAP Atlantique, 2021c).

### 5.4.2 Ria de Aveiro

The Aveiro region is located in the centre of Portugal and integrates 11 municipalities, five of them being coastal areas. The region is characterized by the presence of Ria de Aveiro, an extensive coastal lagoon occupying about 11,000 hectares. It is the most important wetland in northern Portugal, 45 kilometres in length and with a maximum width of 8.5 kilometres (Dias *et al.*, 2000). Due to its importance for nature conservation, diverse protection measures were assigned to Ria de Aveiro. It was classified as an SPA under Directive 79/409/CEE on the conservation of

### Musée des Marais Salants

- Thematic museum, certified as a *Musée de France,* and a research centre with temporary original exhibitions
- Supports salt producers and salt quality labels (e.g. Label Rouge) to strengthen the historic name of *Sel de Guérande.*
- Contributes to the scientific and cultural exchange of the saline Atlantic Arc

### Maison des Paludiers

- Oldest association structure (1971) in the region, dedicated exclusively to salt activity
- Focused on salt collection techniques, the salt worker profession and the interaction between meteorology and salt activity
- Offers guided tours led by former salt workers

### Terre de Sel

- Owned by Cooperative Sel de Guérande, the result of the work between salt workers and naturalists
- Sales, exhibitions, reception information and a place to discover salt pans
- Visits are supervised by salt workers or naturalists, depending on the purpose of the visit and the themes to explore (salt production, gastronomy, ornithology, etc.)

**Fig. 5.2.** Salt pan visitor centres. (Photos courtesy of Buron, 2021.)

wild birds in 1999, and as an SCI under Directive 42/92/CEE (habitats) in 2014, integrating the Natura 2000 network. It is also an Important Bird and Biodiversity Area (IBA).

Despite its importance, the lagoon still faces several environmental and human-caused threats, such as pollution, poor governance and tourism pressures with visible risks to ecological balance (Pita and Silva, 2019). Its salt pans have been abandoned or converted to other uses over time, such as aquaculture. There were only eight traditional salt pans producing salt in 2020 compared with 265 in 1956 (Martins *et al.*, 2020). These salt pans, known as 'Salgado de Aveiro', are an essential component of the history, culture and landscape of Ria de Aveiro's ecosystem. However, only 54 hectares are producing salt, although the overall area of the salt marshes represents approximately 1,243 hectares (Lacerda *et al.*, 2018).

As in Guérande, salt production in Ria de Aveiro uses exclusively artisanal methods. This results in high costs of production and lower competitiveness with salt produced by mechanical means, making it difficult to commercialize the salt produced in the Aveiro salt pans (Silva, 2010). More recently, there has been an increasing interest in recovering salt pans to promote tourism, by combining them with salt production, wellbeing, nature or other recreational activities (Albuquerque *et al.*, 2018; Martins *et al.*, 2020). Salt and salt products are also being commercialized with added value options such as salt with aromatic herbs, *fleur de sel*, salicornia (i.e. a halophyte plant that grows in salt pans and is used in gastronomy) and other salt-based products, such as soap, bath salts and chocolate with *fleur de sel* (Martins *et al.*, 2020).

Both public institutions (e.g. the City Council of Aveiro and the University of Aveiro) and private companies promote the salt pans by offering different experiences – for example:

- salt culture, salt history, salt pan biodiversity;
- aquaculture (bivalves, e.g. oysters);
- artificial beaches for leisure;
- salt water pools for wellbeing activities (i.e. with muds and hypersaline water).

Despite the claim of regional authorities for a specific management entity for the lagoon, the responsibility for managing Ria de Aveiro is still devolved to several public authorities, at different geographical scales. Important decisions regarding the lagoon are still made at a national level, even though these decision-makers are not fully informed of the real situation in the area, which leads to additional difficulties for salt pan entrepreneurs, who must turn to different entities to implement and operate their businesses.

Additionally, some strategic plans and interventions with impacts on the salt pans have been carried out. One of them was the *Polis Litoral Ria de Aveiro –Integrated Requalification and Valorization of the Coastal Border* (Resolution of the Council of Ministers no. 90/2008), which promoted several actions (e.g. preservation and requalification of natural values; creation of small rest and leisure areas and pedestrian circuits in some of the salt pans' protective dams).

In recent decades, the Aveiro salt pans have been part of several regional, national and EU-funded projects, which have contributed to add value to artisanal salt-making and to traditional salt pans. These projects have involved various institutions, such as the University of Aveiro, the Aveiro municipality and the Aveiro Business Association, along with several regional stakeholders.

Among others, these projects have contributed to: the promotion of training courses; recognizing salt as an agricultural product (in the past it was considered as a mining product, which made it impossible to award geographic identification quality labels); promoting a European salt route; promoting business models for salt pans; and recovering and implementing some infrastructure in the salt pans. Nowadays, some of the Aveiro salt pans have their salt certified as an artisanal product, which makes it possible for artisanal salt producers to apply for Protected Designation of Origin (PDO), as for the Portuguese salt producers Tavira, Rio Maior and Castro Marin, who already have a label for their salt and *fleur de sel* (DGADR, 2021, 2021).

The Aveiro municipality has faced an increase in tourism over the last seven years (INE, 2019a), mainly due to boat tours and the proximity to Porto city. The municipality has a population of 78,734 (INE, 2019b) and received, in 2019, 215,338 overnight visitors, which represents an increase of 7% compared with 2018, similar to the national average rate (INE, 2019b). Such rapid tourism growth, especially in the city centre, has created changes in terms of visitor pressure in the urban canals, but also in the salt pans.

The city has the only open-air museum associated with artisanal salt production in the region, a space that allows for both guided and independent visits. The first are possible on request to the education service of the City Museum and the latter by using interpretative panels. Private companies, on the other hand, promote visits to other local salt pans not managed by the City Museum. Even though some visits are scheduled, they do not take into consideration the work hours of salt workers, potentially disturbing them.

### 5.4.3 Guérande and Aveiro: comparative analysis

Comparing both case studies, it is possible to observe the definition of a strategic plan for the salt pans in Guérande since the beginning of its recovery process. It integrated a diversity of stakeholders in a bottom-up participatory strategy approach, whereas in Aveiro individual initiatives took place, mainly led by public institutions and, in recent years, by local private companies, who are the main current exploiters of traditional salt pans in the area, and focus their activity on tourism. There is no specific strategy for artisanal salt-making heritage. The timeline shown in Fig. 5.3 shows key moments in the evolution of both study areas.

Both regions have acquired different key features for traditional salt pans and artisanal salt-making activity due to their evolution over time. Nowadays, Guérande has approximately 350 salt pans producing salt and 300 salt workers, whereas Aveiro has only eight producing salt with 15 salt workers. Although both regions have relatively similar salt marsh areas (i.e. 1343 hectares in Aveiro and 1176 hectares in Guérande), the active salt production area in Aveiro represents only around 7% of its salt marshes (i.e. 54 hectares). In

**Fig. 5.3.** Key moments in both study areas.

Guérande, there are 765 hectares producing salt, which represents around 65% of the area.

This difference has resulted in different production scales, with an average salt annual capacity production of 16,000 tonnes in Guérande and 955 in Aveiro, as well as a difference in the number of active salt pans and salt workers in both areas (Chambre d'Agriculture Loire-Atlantique, 2011; Kortekaas, 2017; INE and DGRM, 2018).

The evolution of tourism activities in the salt pans of both regions has also been different. In Guérande, such activities started to become more evident in the 1970s, with the creation of the first visitor centre exclusively dedicated to salt-making activity, La Maison des Paludiers, followed by two others: Musée des Marais Salants and Terre de Sel. Terre de Sel has had a growing importance in the region since 2000. Since then, the perception of salt workers towards tourism has also changed: it is no longer seen as disruptive to salt production. On the contrary, it has been recognized as a complementary activity that brings awareness and enables the dissemination of natural and cultural heritage associated with salt pans and salt production. In Aveiro, tourism activities in the salt pans started with the creation of the Ecomuseu Marinha da Troncalhada by the city council in 1995. Nevertheless, its activity was initially more associated with environmental education than tourism, promoting study visits to primary and secondary schools. Only over the last ten years, because of private initiatives, has tourism brought a significant dynamic to the salt pans. Moreover, salt pan managers see tourism in this region as the main activity contributing to salt pan survival, which means salt-making is mainly developed as a demonstration process for tourists (Table 5.2).

Comparison between the two case studies allowed us to understand that different governance methods also bring different approaches to the development of specific saltscapes. Guérande already has a specific strategy for tourism activities in salt pans, which is shown by the integrated collaborative and participatory work concerning salt pan governance (cooperative and private owners with the public regional administration), and by the establishment of several partnerships. Such a collaborative approach is missing in Aveiro. The comparison shows that it is important to have an integrated approach, involving all stakeholders and promoting participatory and inclusive governance as an opportunity to develop tourism as a sustainable complementary activity in salt pan regions. The global number of visitors to the salt pan areas of Aveiro is still unknown, which brings difficulties to strategy development. The Aveiro salt pans ought to develop an integrated approach, involving public and private entities, so that the region can maintain its interrelated

**Table 5.2.** Visitor experience in both case studies. (Based on Martins *et al.*, 2020; Buron, 2021; M. Ribeiro, Aveiro, 2021, personal communication.)

| Site | Visitor experience | Rules | Number of visitors in 2019 | Themes explored |
|---|---|---|---|---|
| Guérande | Three visitor centres with guided tours | Limited number of visitors and only in scheduled hours so as not to disturb salt-making activity | 109,422 | Salt production, landscape history, gastronomy, ornithology, fine arts, ethnography |
| Aveiro | Guided tours developed by the Ecomuseum, which integrates guided tours and educational services, and by private companies, as well as independently | No specific rules or scheduled hours regarding. salt-making activity | 13,175* | Salt production, salt history, gastronomy, ornithology Salt pans are used as artificial beaches for leisure and salt-water pools for wellbeing activities (i.e. using muds and hypersaline water) |

*Data for the Ecomuseum only.

natural and cultural heritage and benefit from the sustainable development of tourism.

## 5.5 Concluding Discussions and Future Directions

This chapter has presented different models for preserving and exploring traditional salt pans. Two case studies were analysed, one a well-established salt pan area that offers guidelines on how a sustainable programme, which has been carried out over the years, has contributed to preserving a secular activity, and the other a salt pan area of key relevance in the past, that is mainly used for tourism purposes nowadays.

Some authors have previously identified a set of sustainable activities for the Aveiro salt pans that can be developed to complement salt-making activity (Silva, 2010; Albuquerque *et al.*, 2018; Martins *et al.*, 2020). Martins *et al.* (2020) also refer to the importance of having a management plan for the 'Salgado of Aveiro'. However, this chapter has aimed at discussing whether traditional salt pans can 'survive' with individual initiatives, and whether these initiatives could be used for tourism purposes only, without having an integrative strategy to preserve and develop the secular activity of artisanal salt-making in a sustainable development context, involving a set of stakeholders. From this perspective, the final aim of the chapter is to offer guidelines for traditional salt pans, based on the Guérande case study.

From the literature review, it appears that tourism should be analysed as a sustainable complementary activity in salt pans, contributing to several SDGs. Moreover, UNWTO has proposed 'making tourism a catalyst for positive change' in the context of the International Year of Sustainable Tourism for Development (UNWTO and UNDP, 2017, p. i), considering that it could act as a tool to implement the 2030 Agenda for Sustainable Development, since tourism can contribute to all SDGs, despite being specifically expressed in only three (8, 12 and 14, UNWTO, 2017). In addition, governance is considered a key issue for salt pans, with several authors highlighting its importance in successful strategies for sustainable development (Hall, 2011a, b; Bramwell, 2015).

It is important to considerer how traditional salt pans offer a unique landscape, integrating both natural and cultural heritage, which is an important asset for regions that allows communities to feel a sense of belonging and tourists to experience it. Therefore, sustainable inclusive and participatory practices are still needed to preserve these areas.

The Guérande salt pan case study offers some insights into how cooperation between several stakeholders, with impacts in several areas (such as planning and policies, research, nature conservation, training, tourism, gastronomy and product quality and strengthening) has contributed to a 'three Ps' strategic programme consolidation (Place, People and Product) over the years. Guérande has become a consolidated and a very important part of the region, as it has built social and political awareness about the relevance of salt pans (Kortekaas, 2017).

The following lessons from the Guérande case study are considered of key relevance.

1. To approach salt pans as an indissociable natural and cultural heritage that needs to be preserved by attributing conservation and protection status and measures (Place).
2. To dignify the salt worker profession with certified training in order to transmit this knowledge to the next generation (People).
3. To add value to salt and complementary products (e.g. *fleur de sel*) and activities (e.g. tourism), so that they can be part of rising awareness and the promotion of the natural and cultural heritage associated with salt pans (Product).

Multi-stakeholder involvement over the years is always necessary to effectively implement a sustainable and sustained programme in traditional salt pans. Only with such a participatory, inclusive and consistent approach is it possible to overcome environmental and human-caused threats, such as pollution, poor governance, tourism pressures and others.

In addition, it is also possible to take advantage of market opportunities to empower local communities (e.g. salt workers, tour guides, etc.) – for example, increasing tourist demand, based on experiences related to what visitors perceive as an 'authentic' natural and cultural heritage. Another example would be the growth interest in organic and artisanal products, such as salt, *fleur de sel* and others.

Based on the lessons learned from the Guérande case, we present a schematic outline of the guidelines for traditional salt pans that could be applied to Aveiro (Fig. 5.4).

The findings and guidelines have practical implications for Portugal, since the country has seven municipalities (Aveiro, Figueira da Foz, Alcochete, Olhão, Tavira and Castro Marim – coastal areas – and Rio Maior, an inland area) that try to promote their traditional salt pans as artisanal salt-making and tourism areas (Silva, 2010; Rodrigues *et al.*, 2011), for which the Guérande best practices could be used as a model.

Besides this, Portugal is an emerging tourism destination, whose strategic plan has identified a challenge regarding sustainability that aims 'To ensure the preservation and sustainable economic enhancement of cultural and natural heritage and local identity, as a strategic asset, as well as making this activity compatible with the permanence of the local community' (Turismo de Portugal, 2017, p. 38).

The Guérande best practices could not only impact the country, but also the region, and the 2030 strategy for the Portugal Centre Region identifies some examples of key projects highlighting the relevance of funding nature tourism that preserves landscapes and biodiversity (CCDR-C, 2020). The implementation of these kinds of projects could benefit the Aveiro municipality since traditional salt pans and artisanal salt-making are the 'image'

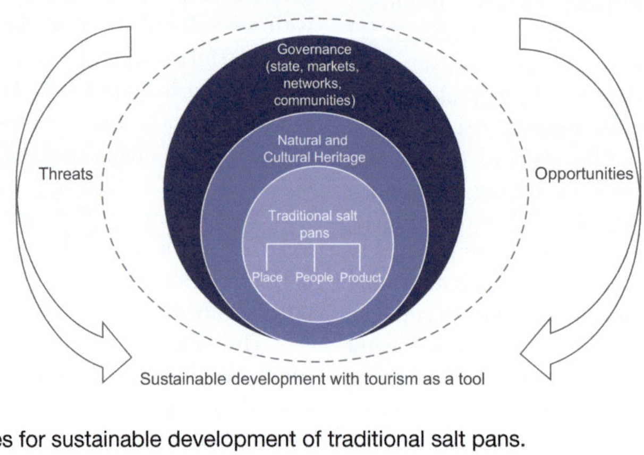


**Fig. 5.4.** Guidelines for sustainable development of traditional salt pans.

of Aveiro and face a serious risk of disappearing or being exclusively used for tourism purposes, reducing salt-making to a mere tourist demonstration.

Visiting salt pans is an outdoor experience that does not allow for the simultaneous presence of a large number of people, since salt pans are natural areas with protection status. In that sense, the development of tourism in salt pans is an opportunity for the local economy in a pandemic context, and the current situation has shown that these areas offer proper visiting conditions, making them an interesting alternative for visitors.

This study also provides some insights regarding sustainable strategies for traditional salt pans that could be useful for researchers, policymakers and other stakeholders. However, it has several limitations due to its restricted scope and aptitude for generalization (Yin, 2013). In the future, this study could be used as a starting point to explore how the implementation of Agenda 2030 and the 17 SDGs can be an opportunity to consider the urgency of preserving the natural and cultural heritage of traditional salt pans, and use tourism as a tool to do this.

## Acknowledgements

The first author of this work was supported by the Portuguese Foundatio for Science and Technology (FCT) with a PhD grant (SFRH/BD/145485/2019), which was co-funded by FSE, through the Centro 2020 programme.

## References

Albuquerque, H., Silva, A.M., Martins, F. and Costa, C. (2018) Wellness tourism as complementary activity in salt pans regeneration. In: *Azara et al. Tourism, Health, Wellbeing and Protected Areas*. CAB International, Wallingford, UK, pp. 56–68.

Amorim, I. (2001) *Aveiro and the salt route, from production to consumption* [15th to 20th centuries], Edited by Câmara Municipal de Aveiro. Aveiro, Portugal.

Asmelash, A.G. and Kumar, S. (2019) Assessing progress of tourism sustainability: Developing and validating sustainability indicators. *Tourism Management* 71, 67–83. DOI: 10.1016/j.tourman.2018.09.020.

Bastos, M.R. (2009) On the track of salt: Adding value to the history of saltponds exploration in the coastal management scene of Aveiro lagoon. *Journal of Integrated Coastal Zone Management* 9(3), 25–43.

Boluk, K.A., Cavaliere, C.T. and Higgins-Desbiolles, F. (2019) A critical framework for interrogating the United Nations Sustainable Development Goals 2030 Agenda in tourism. *Journal of Sustainable Tourism* 27(7), 847–864.

Bramwell, B. (2015) Theoretical activity in sustainable tourism research. *Annals of Tourism Research* 54, 204–218.

Bramwell, B., Higham, J., Lane, B. and Miller, G. (2017) Twenty- five years of sustainable tourism and the Journal of Sustainable Tourism: looking back and moving forward. *Journal of Sustainable Tourism* 25(1), 1–9. DOI: 10.1080/09669582.2017.1251689.

Buron, G. (2021) Modalités de transmission d'un patrimoine immatériel : Le cas de savoir-faire des marais salants de Guérande • Batz • baie de Mesquer. *Colloque international le sel : histoire – patrimoine – tourisme, travail, savoir-faire et mémoires ouvrières*, Arc-et-Senans, 05-06 avril 2018.

CAP Atlantique (2013) Étude Poids économique du tourisme, Leaflet.

CAP Atlantique (2021a) La dynamique maritime, la saliculture. Available at: https://www.cap-atlantique.fr/entreprendre/la-dynamique-maritime/la-saliculture (accessed 20 February 2021).

CAP Atlantique (2021b) Les filières d'excellence, le tourisme. Available at: https://www.cap-atlantique.fr/entreprendre/un-territoire-attractif/le-tourisme (accessed 20 February 2021).

CAP Atlantique (2021c) Les économies primaires, la saliculture. Available at: https://www.cap-atlantique.fr/linstitution/les-politiques-publiques-de-cap-atlantique/les-economies-primaires/la-saliculture (accessed 20 February 2021).

Carlisle, S., Kunc, M., Jones, E. and Tiffin, S. (2013) Supporting innovation for tourism development through multi-stakeholder approaches: Experiences from Africa. *Tourism Management* 35, 59–69.

Chambre d'Agriculture Loire-Atlantique (2011) *Diagnostic salicole avec la participation du Comité Professionnel Salicole*. CAP Atlantique, Nantes.

Chhabra, D., Healy, R. and Sills, E. (2003) Staged authenticity and heritage tourism. *Annals of Tourism Research* 30(3), 702–719.

Chylińska, D. (2019) Lighthouses as traditional coastal landscape heritage and the basis for lighthouse tourism development: the case of Poland. *Journal of Tourism and Cultural Change* 19(3), 315–344. DOI: 10.1080/14766825.2019.1700991.

CCDR-C (2020) *Strategic Vision for the Centre Region 2030*. Available at: http://www.ccdrc.pt/index.php?option=com_docman&view=download&alias=5300-visao-estrategica-regiao-centro-2030&category_slug=2020&Itemid=739 (accessed 8 January 2021).

DGADR (2021) Traditional Portuguese Products. Available at: https://tradicional.dgadr.gov.pt/pt/produtos-por- regime-de-qualidade/dop-denominacao-de-origem-protegida (accessed 20 January 2021).

Dias, G. (2005) SALT and its ambivalent dimension: taste of food and symbol of religious preservation. In: *Proceedings of 1st International Seminar on Portuguese Salt*. University of Porto, Porto.

Dias, J.M., Lopes, J.F. and Dekeyser, I. (2000) Tidal propagation in ria de aveiro lagoon, portugal. *Physics and Chemistry of the Earth* 4(25), 369–374.

Domínguez-Quintero, A.M., González-Rodríguez, M.R. and Paddison, B. (2020) The mediating role of experience quality on authenticity and satisfaction in the context of cultural-heritage tourism. *Current Issues in Tourism*, 248–260.

EC - Publications Office of the European (2005) Nature 2000: conservation in partnership. Available at: https://ec.europa.eu/ (accessed 20 January 2021).

Hall, C.M. (2011a) A typology of governance and its implications for tourism policy analysis. *Journal of Sustainable Tourism* 19(4–5), 437–457.

Hall, C.M. (2011b) Policy learning and policy failure in sustainable tourism governance: from first- and second-order to third-order change? *Journal of Sustainable Tourism* 19(4–5), 649–671.

Higgins-Desbiolles, F. (2018) Sustainable tourism: Sustaining tourism or something more? *Tourism Management Perspectives* 25, 157–160.

Hocquet, J.C. (1982) *Le Sel et la Fortune de Venise*, Vol. 1. Production et Monopole. Edited by Presses del'Université de Lille III, Lille, France.

Idowu, O. (2016) Criticisms, constraints and constructions of case study research strategy. *Asian Journal of Business and Management* 5(4), 2321–2802.

INE (2019a) Guests (No.) in tourist accommodation establishments by Geographic localization (NUTS – 2013) and Type (tourist accommodation establishment); Annual. Available at: www.ine.pt (accessed 8 January 2021).

INE (2019b) Resident population (No.) by Place of residence (NUTS – 2013), Sex and Age group; Annual, INE, Annual Provisional Resident Population Estimates. Available at: www.ine.pt (accessed 8 January 2021).

INE and DGRM (2018) Fisheries Statistics 2017. Available at: www.ine.pt (accessed 8 January 2021).

Jamal, T. and Stronza, A. (2009) Collaboration theory and tourism practice in protected areas: Stakeholders, structuring and sustainability. *Journal of Sustainable Tourism* 17(2), 169–189.

Katelieva, M., Muhar, A. and Penker, M. (2019) Nature-related knowledge as intangible cultural heritage: safeguarding and tourism utilisation in Austria. *Journal of Tourism and Cultural Change* 18(6), 673–689. DOI: 10.1080/14766825.2019.1693581.

Kim, S., Whitford, M. and Arcodia, C. (2019) Development of intangible cultural heritage as a sustainable tourism resource: the intangible cultural heritage practitioners' perspectives'. *Journal of Heritage Tourism* 14(5-6), 422–435. DOI: 10.1080/1743873X.2018.1561703.

Korovessis, N.A. and Lekkas, T.D. (2009) Solar saltworks' wetland function. *Global Nest Journal* 11(1), 49–57.

Kortekaas, K. (2017) Salt in our veins. The patrimonialization processes of artisanal salt and saltscapes in Europe and their contribution to local development. PhD thesis. University of Barcelona, Barcelona, Spain.

Kortekaas, K.H. (2004) Sustainable tourism initiatives in European saltscapes. *Sustainable Tourism* 9, 199–207.

Kyvelou, S.S. and Gourgiotis, A. (2019) Landscape as connecting link of nature and culture: Spatial planning policy implications in Greece. *Urban Science* 3(3), 81. DOI: 10.3390/urbansci3030081.

Lacerda, C., Morais, L., Dinis, C. and Santos, J. (2018) Salt of Aveiro project – Market study and business opportunities. Available at: https://saldeaveiro.pt/estudos-e-publicacoes/ (accessed 10 December 2020).

Le Drogo, G. (2006) Étude de patrimonialisation des sites français. Interreg III SEL Sel de l'Atlantique (Unpublished report).

Le Guerandais (2021) Organisation, operation and figures. Available at: https://www.leguerandais.fr/en/cooperative/organisation-operation-and-figures (accessed 20 February 2021).

Martins, F., Silva, A.M., Albuquerque, H.*et al.* (2013) Sustainability of salt pans – new approaches, new activities. In: Borrego, C. (ed.) *Rethinking the Environment: Luxury or Inevitability? XII CENA (National Congress of Environmental Engineering)*. Department of Environment and Planning – University of Aveiro, pp. 590–595.

Martins, F., Pedrosa, A., da Silva, M.F., Fidélis, T., Antunes, M. *et al.* (2020) Promoting tourism businesses for "Salgado de Aveiro" rehabilitation. *Journal of Outdoor Recreation and Tourism* 29, 100236. DOI: 10.1016/j.jort.2019.100236.

Ménanteau, L. and Nedelec, L. (2007) Projet SAL. Action 5A. Interreg III B Project SAL Sel de l'Atlantique (Unpublished report).

Pafi, M., Flannery, W. and Murtagh, B. (2020) Coastal tourism, market segmentation and contested landscapes. *Marine Policy* 121, 104189.

Pita, C. and Silva, A.M. (2019) Identify and analyse gaps between expert knowledge and social representations of CH at risk, PERICLES PrEseRvIng and sustainably governing Cultural heritage and Landscapes in European coastal research report. Available at: https://www.pericles-heritage.eu/wp-content/uploads/2020/04/PERICLES_D4.2.pdf (accessed 2 December 2020).

Postma, A. and Schmuecker, D. (2017) Understanding and overcoming negative impacts of tourism in city destinations: conceptual model and strategic framework. *Journal of Tourism Futures* 3(2), 144–156.

Ramsar and UNWTO (2012) Destination wetlands: Supporting sustainable tourism. In: *Secretariat of the Ramsar convention on wetlands*. Gland, Switzerland, and World Tourism Organization (UNWTO), Madrid, Spain.

Ramsar Convention Secretariat (2016) *An Introduction to the Convention on Wetlands*, 7th edn. (previously The Ramsar Convention Manual). 107. Gland, Switzerland.

Ritz, E. (1996) The history of salt – Aspects of interest to the nephrologist. *Nephrology Dialysis Transplantation* 11(6), 969–975. DOI: 10.1093/oxfordjournals.ndt.a027517.

Rodrigues, C.M., Bio, A., Amat, F. and Vieira, N. (2011) Artisanal salt production in Aveiro/Portugal – an ecofriendly process. *Saline Systems* 7(3), 14. DOI: 10.1186/1746-1448-7-3.

Sainz-López, N. (2017) Comparative analysis of traditional solar saltworks and other economic activities in a Portuguese protected estuary. *Boletin de Investigaciones Marinas y Costeras* 46(1), 171–189.

Siakwah, P., Musavengane, R. and Leonard, L. (2020) Tourism governance and attainment of the sustainable development goals in Africa. *Tourism Planning and Development* 17(4), 355–383. DOI: 10.1080/21568316.2019.1600160.

Silva, A. (2010) Estratégia para uma Gestão Sustentável do Salgado de Aveiro (Strategy for a Sustainable Management of Salgado de Aveiro). Master Thesis. University of Aveiro, Aveiro, Portugal.

Turismo de Portugal (2017) Tourism strategy 2027. Available at: http://estrategia.turismodeportugal.pt/sites/default/files/Estrategia_Turismo_2027_TdP.pdf (accessed 10 December 2020).

Univers-sel (2021) Une association de solidarité internationale. Available at: http://www.universsel.org/ (accessed 8 January 2021).

UNWTO (2017) Press release: A roadmap towards 2030: The legacy of the International Year of Sustainable Tourism for Development 2017. PR 17137, 19 Dec 17. Madrid.

UNWTO (2018) *Tourism and Culture Synergies, Tourism and Culture Synergies*. Madrid.

UNWTO (2021) Sustainable development. Available at: https://www.unwto.org/sustainable-development (accessed 8 January 2021).

UNWTO and UNDP (2017) Tourism and the Sustainable Development Goals – Journey to 2030, Tourism and the Sustainable Development Goals – Journey to 2030.

Ville de Guérande (2021) Données clés. Available at: https://www.ville-guerande.fr/mairie-vie-municipale/decouvrir-la-ville/presentation-de-la-ville (accessed 2 February 2021).

Waligo, V.M., Clarke, J. and Hawkins, R. (2012) Implementing sustainable tourism: A multi-stakeholder involvement management framework. *Tourism Management*, 342–353.

Xie, P.F., Lee, M.Y. and Wong, J.W.C. (2020) Assessing community attitudes toward industrial heritage tourism development. *Journal of Tourism and Cultural Change* 18(3), 237–251.

Yin, R.K. (2013) *Case Study Research: Design and Methods*. Sage Publications, Thousand Oaks, USA.

# 6 Strategic Considerations for Sustainable Tourism Development of the Micro-destination East Belgium

**Laura Rauschen[1], Anastasia Traskevich[2]* and Martin Fontanari[3]**
*[1]Hochschule Fresenius University of Applied Sciences, Idstein, Germany; [2]Belarusian State Economic University, Minsk, Belarus; [3]International School of Management (ISM), Cologne, Germany*

## 6.1 Introduction

After years of mass tourism, more and more tourists are looking for alternatives and niche markets. Managerial approaches to regional development through tourism always concentrate on a destination as a regional unit. However, hardly any attention has been paid to the hierarchy of destinations as the objects for regional management and governance, particularly to the lowest territorial level in this hierarchy – the level of a *micro-destination* (Papatheodorou, 2006; Stancioiu *et al.*, 2011; Segota, 2018). Therefore, at the current stage of academic and professional discourse, even the precise and universal definition of a micro-destination presents a clear research gap.

From this theoretical perspective, the current research is focused on the practical example of tourism development in East Belgium with the hypothesis that this emerging tourist destination can be evaluated and further managed sustainably as a micro-destination. East Belgium is an autonomous region of Belgium that consists of a German part and a French part. It is a small region that differs strongly from the neighbouring territories.

The unique and authentic natural, cultural and social characteristics of this autonomous area create substantial prerequisites for tourism development which are, however, too diverse from the larger territorial entities and from the overall offering and tourist image of Belgium as a whole country. The diversity, authenticity and uniqueness of this autonomous region substantiate the thesis of the present research that the optimal approaches to destination management and governance can be elaborated for tourism development in East Belgium when they are designed and implemented on the most locally-focused level – thus, the level of a micro-destination.

Therefore, the current study aims to elaborate on an integrative definition of a micro-destination, with applied focus to implement a locally-based approach to build on the tourism development strategy and related product policy of the micro-destination East Belgium. The achievement of the research goal can present a best practice case study on the implementation of strategic decisions at the level of a micro-destination because the strongest local focus is required due to the diverging attributes of the neighbouring regions, on the

*Corresponding author: anastasia.traskevich@akalem.de

 *Tourism Planning and Development in Western Europe* (eds K. Andriotis *et al.*)
DOI: 10.1079/9781800620797.0006

whole and particularly in terms of their tourism development.

Thus, the expected applied impact of defining a distinctive and authentic region as a micro-destination lies in elaborating on the specific methodological approach to tourism development strategy. This approach presents scientific, methodological and practical novelty in terms of its focus on intensive involvement of local tourism resources, local community, policymakers and stakeholder representatives for integrative participation in the development of creative strategic solutions, which are precisely targeted for their native region. From this perspective, the level of a micro-destination appears the most relevant level of regionally-based strategic thinking to ensure sustainable and resilient-oriented destination development.

The empirical part of the research presents an expert interview and workshop, both targeted to the implementation level of a micro-destination. In addition, the direct practical outcomes of the workshop have been examined for the case of the micro-destination East Belgium and briefly presented in this chapter. The expected advantage of the proposed focus on the level of a micro-destination is that (i) elaborated tourism development projects create benefit within the authentic, unique but unexploited local tourism potentials in the region; and (ii) the expert workshop conducted at the targeted micro-level helps to identify and describe tourism themes, tourism products and development projects with the precise and detailed involvement of local expert knowledge and local stakeholder interests.

## 6.2 Defining a Micro-destination

While there has been a focus in research circles on destinations and how to manage them, little research has been conducted on the specific area of micro-destinations. Existing literature has not provided one universal definition for the term micro-destination. Different authors (Papatheodorou, 2006; Andrades and Dimanche, 2015; Camillo *et al.*, 2015; Segota, 2018) define it as a narrow geographic area of a country, merely defined through the criteria of a portfolio of tourism products that are available in the location. From this perspective, a city is most often considered as a typical example of a micro-destination. The concept of a micro-destination has, thus, so far been implemented mainly through a marketing campaign aimed at the unique and autonomous tourism reputation of a touristic city. Moreover, the majority of existing definitions portray the landscape and scenery of micro-destinations (e.g. Croatia tourism). It has been much less frequent, as yet, to feature the uniqueness of the people who create the indigenous community of a micro-destination. It can be assumed, therefore, that each author uses the term micro-destination subjectively and according to their further applied interests: there is no universal definition for the term micro-destination, as well as no universally comprehensive components and attributes to define this term.

The most specific and developed definition of the term micro-destination has been provided by Hernández-Martín *et al.* (2016, p. 775). In their paper, the concept of micro-destination is defined as follows:

> A small geographical unit that is highly dependent on tourism. It comprises a wide range of tourism facilities and has a differentiated image and tourism typology (tourism products). Besides, a micro-destination is a useful individual unit for decision-making in tourism management and planning. For a more operational definition, a micro-destination is a spatial unit of statistical analysis characterized by a high density of establishments of tourism characteristic industries, homogeneous statistical tourism information, and a spatial continuity.

The disadvantage of this definition is, however, that it does not define quantitatively what can be precisely considered as the high density of tourism accommodations. There is no indicator or a minimum number of tourism accommodations that a region must have in order to fit into the definition.

From the perspective of the current research, the first important criteria to define a micro-destination is the density of tourism establishments. Besides, the size and region-focused characteristics of a micro-destination are important to differentiate a separate tourism object (like a hotel) from a micro-destination as specific regional unit. In this context, existing definitions of a micro-destination lack the following attributes. There is no indication of a minimum or maximum size of a micro-destination. In addition, there is lack of qualitative criteria used in previous research to emphasize the regional context of a micro-destination. No clear characteristics mark the difference between different hierarchical types of a destination: micro-level or meso-level, etc. Furthermore, there is currently no focus on local stakeholders and the local community, nor any details about the regional economic development and sustainable growth through tourism activities. In addition, destinations are identified as micro-destinations but are not marketed as such – therefore, the term micro-destination remains isolated from real-life tourism practice and expert use. All the aforementioned gaps in conceptualizing the term micro-destination comprise, therefore, the criteria on which the present authors' definition of a micro-destination has been elaborated. This author integrative definition is targeted to the object of the current research (East Belgium) and is formulated as follows:

> A stable and dynamic location which is distinctive from neighbouring regions, both historically and culturally, as well as in terms of its natural environment; and gains tourism potentials through these aspects of local diversity and uniqueness. Within the same region, local community and tourism stakeholders form a homogenous unit by sharing similar values, rules, and culture.
> (Rauschen, 2020)

The key characteristic of the presented definition, which distinguishes it from previous ones, is the precise regional focus – which is strongly related to the empirical context of the current research. According to the integrative definition of a micro-destination in this research, the following criteria of defining a micro-destination are introduced. Firstly, the micro-destination differentiates itself from neighbouring regions not merely politically, administratively and geographically but mainly culturally, naturally and historically. Secondly, the community within the micro-destination shares the same values and rituals and has the same education system. Thirdly, in the micro-destination, they speak another language or dialect than in neighbouring regions. Lastly, the micro-destination could be viewed as a minority within the country through its natural or cultural boundary to other regions. If the majority of these criteria are fulfilled, then the region can call itself a micro-destination and market itself as such.

From this perspective, East Belgium is defined as a micro-destination based on the following factors, which confirm the aforementioned criteria. As an independent destination, each municipality of East Belgium is too small on the travel market. However, the establishment of the three different parts of the micro-destination – Eifel, Hohes Venn and Eupener Land – enables the 'Tourismusagentur Ostbelgien' (TAO) (the local destination marketing organization, DMO) – to market the different landscapes and 'trump cards' as a region of versatile nature experience. This requires cross-community cooperation of the actors. Historical, cultural and natural values, and how they can be combined to form a homogeneous tourism product, are important attributes in the tourism development of East Belgium. Furthermore, locally specific core values in their connection with the improvement of sustainability of the region and the wellbeing of local communities play an important role. Local communities and stakeholders, with their ability to elaborate a homogenous, attractive and unique tourism product with the help of the destination label 'Made in Ostbelgien' (MiO) help to identify the region as a unique one.

However, the components and attributes which make the term micro-destination relevant for the applied case of East Belgium are not solely the uniqueness of the history, nature and cultural values of the region. Application of the concept of a micro-destination for the autonomous region East Belgium also contributes to the issues of sustainability in tourism development in this region, particularly the cases of involvement of local natural and cultural heritage. In this sense, the concept of a micro-destination strategically emphasizes constant investment in the conservation and development of local natural and cultural heritage – this is to make sure that there is enough reason for tourists to come to the micro-destination over the years. Besides, another important factor regarding sustainability within the application of the concept of a micro-destination is the integration of local stakeholders and the local community within the region, as only their partnership allows a micro-destination to turn into something unique, holistic and emergent. This locally-based partnership comprises, thus, the societal aspect of the application of the concept micro-destination for the goals of sustainable tourism development in the region.

## 6.3 Tourism Potentials of East Belgium as a Micro-destination

The micro-destination East Belgium consists of two different parts: the German-speaking community and the French-speaking community. The territory of the German-speaking community of Belgium covers 854 square kilometres. It lies in the east of the country on the borders with the Netherlands, Germany and the Grand Duchy of Luxembourg. Between the two parts of the German-speaking community lies a landscape which is unique in Europe: the High Fens. This upland moor is the last one in Europe and is part of the protected natural reserve Hohes Venn-Eifel. This nature park stretches between Eupen in the north and Burg-Reuland in the south and is partly on the territory of French-speaking communities, namely Malmedy and Weismes. Together with the Nordeifel Nature Park, the Hohes Venn-Eifel Nature Park forms the 2439.7 square kilometres of the German-Belgian Nature Park (Ostbelgienlive). The French-speaking community consists of the two municipalities Malmedy and Weismes. They also belong to the label of East Belgium even though the inhabitants speak French.

The main public stakeholder in the field of tourism is the domestic Tourism Agency and DMO of East Belgium called 'TAO'. Since 2011, it has been responsible for the product development and internal and external marketing of East Belgium (Schmitz, 2011). It is also responsible for the establishment of the project *Tourism Development East Belgium 2025 Mission Statement*. This official document, together with the document *The Regional Development Concept of the German-speaking Community* (REK), Volumes 1–5, *inter alia* define strategically the current tourism situation in the micro-destination East Belgium.

To define East Belgium as a micro-destination, the analysis of tourism potentials of the region has been made with regard to the aforementioned criteria, with precise focus on the unique and diverging local natural, historical and political characteristics of the region. Out of this analysis, it turns out that in this case there are equally strengths and weaknesses for tourism development at the level of a micro-destination. However, opportunities are more present than threats. The strengths are a summary of everything that has been mentioned before in this chapter: it is all about culture, nature and language as well as regional products and local stakeholders. In fact, the strengths are created by the attributes that mark East Belgium as a micro-destination according to the region-focused definition presented in the current research.

The weakness with the most important consequence is the lack of data necessary to provide an offer which is adequate for the 'profile tourist' and which can be easily changed and adapted to current market trends. This means that quantitative and qualitative market data are missing in the application of strategic thinking in tourism development at the micro-level – the way that the present concept of a micro-destination suggests. This is required in order to exactly know the needs and preferences of targeted tourist groups and to elaborate on the niche tourism themes which would be most relevant to the unique preconditions of the micro-destination. Also, the fact that East Belgium and its attractions are not marketed internationally to differentiate this autonomous region from the rest of the country can add to the reasons why the present concept of a micro-destination is relevant to ensure a boost in domestic and, especially, international tourism in East Belgium.

Furthermore, most of the threats are general ones and could be present in every destination at any hierarchical level, such as monocultures in agriculture, family-run businesses that cannot be maintained in the next generation, or a shortage of professional workers in the tourism sector. In contrast, the opportunities for the micro-destination East Belgium lie in a targeted marketing concept that profiles East Belgium as an exciting tourist destination offering unique and authentic 'local experiences'. There are 'Lighthouse' projects that have already been launched by the TAO, such as the 'Stoneman' and long-distance hiking trails. A crucial opportunity is the sustainable tourism, so ensuring environmentally, socially and economically compatible tourism.

Concerning tourism attractions in the region, there are different important points to mention. Firstly, East Belgium is considered the 'mountain region' or the 'roof of Belgium', as nature is hilly in contrast to the rest of the country. In spring, there are many narcissi meadows, which create a unique natural spectacle. In contrast, in autumn, the forests and their inhabitants become more important, as the region is known for the rutting of the stags. The main all-year attraction is, undoubtedly, the unique moor plateau of the High Fens, which is inhabited by rare species of moorland and wetland plants and endangered animal species such as the black grouse (Ostbelgien, 2018).

Hiking is one of the main tourism products in the micro-destination East Belgium. The honeycomb-hiking network, which was created and opened in 2019 for hikers, also leads through the High Fens. There are more than 50 different routes available depending on what tourists prefer: long routes, routes with kids and routes with many elevation metres (Ostbelgien, 2018). Biking is another important tourism product. There is the well-developed network of the Ravel cycle paths and the Vennbahn (more than 125 kilometres), which are a good option for pleasurable Sunday rides with kids; and there are challenging routes for professional sportsmen (Ostbelgien, 2019). The Vennbahn is cross-bordering with Germany and Luxembourg. In addition, the sport, leisure and tourist Centre Worriken and the Weser Reservoir Visitor Centre contribute to a diverse sports tourist offer in East Belgium (Heukemes, 2009). There is also an Action & Fun Center in Grüfflingen with a cart track, a bowling centre, a railbike and a 7D cinema. Furthermore, regarding cultural heritage, the micro-destination East Belgium has other touristic places to offer: ruins of old castles, museums, historical centres, slate mines and guided tours through factories that produce regional products (Ostbelgien, 2019). In addition, there are different traditional festivals and events, which are highly appreciated by tourists, such as the carnival folklore.

Gastronomy is certainly an essential tourism product. The food in East Belgium can be compared to German food in its quantity and French food in its quality, which is a good mix for all different kinds of tourists. Furthermore, many restaurants offer local, regional products, which present an additional asset of the gastronomic experience in the micro-destination. Considering regional products, one has to mention beer, chocolate, pralines, waffles and fries, as is the case for the rest of Belgium. However, in East Belgium, they also produce other regional, high-quality products, for which they created the independent MiO. The label 'Made in Ostbelgien' stands for regional quality products that reach the consumer via direct routes. These products include honey, Ardennes ham, game pâté, schnapps, sausages, bread, coffee, beer, milk and yoghurt, and mustard (Ostbelgien, 2018). Thus, in fact, the main tourism potential of the destination lies in the development of locally-based tourism projects, which involve gastronomy, cultural and natural heritage, and hiking and biking activities. These attributes mark the difference between the micro-destination East Belgium and the surrounding territories and can, at the same time, be viewed as the strengths of East Belgium as a micro-destination.

Finally, the following labels are solely available in East Belgium. The aforementioned MiO label signifies 'Made in Ostbelgien' and is a quality label for products that are produced or made in East Belgium. Indicating this label on the menu, restaurants show their clients that they work with regional products. The label ensures that the restaurant is officially quality-proved and has the high service standards of East Belgium (Ostbelgientourismus, 2019). Another label is the 'bed+bike' label, which comes from Germany and has been adapted to East Belgium. The label indicates that the accommodation is bike-friendly and situated next to the Ravel bike path (Ostbelgientourismus, 2019). These labels can help to promote East Belgium as a micro-destination and create a benefit to the region by indicating unique selling propositions (USPs). You can find them nowhere else.

## 6.4 Methodological Approach to Development Strategy and Product Design in the Micro-destination East Belgium

The basic characteristic of the present methodological approach is the focus on tourism themes, which have been introduced to develop a new strategy and new product designs. The related strategic considerations are to be developed with the aim of increasing tourism demand, to improve the marketing, branding and image of the destination, and to contribute to the sustainable development of local communities. Thus, the current methodological approach to development strategy and product design in the micro-destination covers two successive structural parts: firstly, an expert interview with the representative of the local tourism administration is undertaken to gain the key expert insights on tourism development in East Belgium; and secondly, a workshop with the integrative participation of the local communities is conducted to enhance creative design approaches to product development in the micro-destination. Thus, the key characteristic of the present methodological approach is its objective to get in contact with local experts and community representatives to develop sustainable tourism themes and USPs for the micro-destination East Belgium.

Within the first step of the methodological approach to development strategy and product design for the micro-destination, a brand pyramid is introduced. The implementation of the pyramid is built on the results approbated by Partale and Traskevich (2018) for five experimental micro-destinations in Belarus. The brand pyramid for the micro-destination East Belgium is based on an in-depth analysis of the region and summarizes the envisaged marketing of East Belgium as a regional tourism brand (Fig. 6.1).

The brand pyramid combines the three core values of the micro-destination East Belgium: nature, culture and gastronomy. The pyramid gives an overview of the brand positioning in the national and international market and

**Fig. 6.1.** Positioning and branding pyramid of the micro-destination East Belgium. (Authors' illustration.)

highlights the emotional benefits that tourists experience when visiting the region. At the same time, the applied part of this research aims to develop further the rational attributes of this brand pyramid, which then permits the creation of innovative tourism themes for the micro-destination East Belgium. For the practical aims of the current research, the relevance of the presented pyramid lies in its links to the diverse benefits that tourists gain while visiting East Belgium. This experiential presentation allowed us to explain the empirical research benefits of the present conceptual approach to the audience, thus ensuring its practical approbation. In terms of the implementation, the content of the pyramid also transfers to the results of the expert interview. Moreover, through the pyramid, these results could be clearly interpreted for the participants of the workshop. Thus, the pyramid also acts as a bridging tool between the two structural parts of the present empirical study.

The expert interview was conducted with the local destination management organization, TAO. This interview was scheduled for 30 minutes and was digitally recorded. It aimed to identify the key strategic questions of tourism development in East Belgium (based on the existing strategic document *Tourism Development East Belgium 2025 Mission Statement*) and to correlate them with the authors' conceptual presentation of East Belgium as a micro-destination.

The interview is structured within the focus blocks defining the target groups which are specific to the micro-destination East Belgium as well as the related strategies to attract the relevant targets for the micro-destination (sustainability component; USPs in the region which can define the region itself as a micro-destination; marketing strategies to develop the reputation of East Belgium as a unique and autonomous micro-destination). The purpose of the interview was accomplished in terms of successfully gaining an overall insight of East Belgium as a micro-destination from a professional (practical) perspective. The information created a base for further design of the workshop, and in particular allowed for further development of the two business games. At the same time, the conceptual considerations of the present study regarding the term 'micro-destination' were confirmed empirically during the expert interview: the results of the interview provided outcomes which were clearly differentiated from the neighbouring regions in Belgium and proved the relevance of the term to be practically applicable to East Belgium.

Regarding the results of the interview, the second part was elaborated and carried out within the current methodological approach in the form of a workshop with local stakeholders and community representatives. This aimed to elaborate sustainable, creative tourism themes that allow the promotion of natural and cultural heritage and that are beneficial for the local community. The intended participants were selected carefully from experts in different tourism fields, such as hotel management and tourism office management as well as employees at tourist facilities and representatives of the local community. However, the actual time arranged for the workshop (April 2020) only allowed a very limited participation rate (three participants), due to Covid-19 and the limited organizational opportunities for alternative modes of professional workshops and discussions at the beginning of the lockdowns in Western Europe.

The duration of the whole workshop was about 90 minutes. At the beginning of the workshop, the term 'micro-destination' was explained with the aim of gaining the critical opinion of experts. For the workshop, the methodology of Partale and Traskevich (2018) was used. Firstly, basic themes, which represent the main tourist core markets, were identified. These do not have a clear competitive advantage and could easily be found in other regions as well. Three basic themes had already been identified for the micro-destination East Belgium during the market analysis, namely nature tourism, cultural tourism and gastronomic tourism.

Then, trend themes were identified. They represent markets that have growth potential and can develop a limited use of competitive advantage (Partale and Traskevich, 2018). Three trend themes had already been identified for the micro-destination East Belgium on the basis of the expert interview, namely hiking, cycling and gastronomy.

The most important themes for the level of a micro-destination are, undoubtedly, the USP themes. They are based on the competitive positioning characteristics of the destination. In addition, particularly for the level of a micro-destination, they allow the region to differentiate itself from other destinations with competitive advantages (Partale and Traskevich, 2018). According to the *Tourism Development East Belgium 2025 Mission Statement* (TAO Beratungs- und Management GMBH, 2013, p. 40), the USPs of the micro-destination East Belgium are as follows: 'Individual hiking success stories such as those of a very successful bog hiking offer and a successful culinary hiking offer show how a 'Unique Selling Proposition', a unique selling position can be achieved with the landscape, the cuisine, the language competence and extraordinary achievements.'

Then, two business games were planned with the aim of promoting creative design thinking for the development potential of the micro-destination East Belgium. They were focused on the experts' own critical experience in the tourism sector and aimed to get creative ideas on paper and then to discuss these experiences.

The first business game was called 'brain writing' and consists of silently building collective ideas. It has three purposes:

- to produce ideas and stimulate creativity in product design and implementation;
- to overcome blocks in terms of regional development and think critically about the issue of sustainable tourism development in the micro-destination;
- to plan approaches for strategy implementation and to solve ongoing problems of tourism development.

The procedure is the following for the 'brain writing' business game. First, tourism themes were announced and written down. Subtopics were then added to the main tourism themes. Then, every participant silently wrote his or her ideas down on a piece of paper. After three to five minutes, the papers were exchanged between the participants and read, prompting them to add other ideas. For this research, two main tourism themes were identified for the micro-destination East Belgium: bog hiking and culinary hiking. The same subtopics were offered for these tourism themes: existing offers and missing offers. If participants had some other ideas related to the topic, they were free to add anything they liked. As the participants exchanged their sheets, it was not possible to further differentiate and compare the answers between different stakeholders. However, the beneficial outcome of this game was that it generated overall and collective ideas: one idea helped the next idea to grow and develop. The most important empirical results achieved in this game were the following.

- *High Fens:* the experts stated that products and tours dedicated to families and kids are missing. There are also no routes available where pets are allowed. Team-building and related activities such as survival camps are not offered. Another problem is the lack of parking. There is also no official, safe bike route that goes from the Eifel to the High Fens. In the High Fens, there is a youth hostel, but no universal tourist accommodation.
- *Worriken:* there is a range of different accommodations for different target groups. Different outdoor activities are also available, for individuals and groups. What is missing are indoor activities and offers in case of bad weather. Furthermore, there are no offers like sports courses for overnight guests although the infrastructure is perfect to offer some courses such as water gymnastics, Zumba or yoga.
- *Hiking and biking in combination with gastronomy:* the infrastructure has improved considerably since 2019. There are different offers available such as hiking and biking without luggage. What is missing is the focus on gastronomy, products that relate both the active and the culinary parts. There are only a few options available that focus on families and children.

Once this first business game was completed in the workshop, the second game, called 'persona creation' was explained to the participants. In this exercise, the participants created personas and identified their archetypical behaviour, characteristics and motivations. Based on the essence of the approach developed by Schoop (2019), the following aspects needed to be identified for this game:

- psychographic data: name, age, marital status;
- background: history, expectations and experiences of the person;
- behaviour: how does the person behave in specific tourism-related situations?
- aims: what intentions, motivations and goals (private/professional) does the person have?

Each participant chose a persona from a different target group. For this purpose, the demand groups used were those identified during the first steps of the current study. Within the present methodological approach, the target group of the chosen persona was selected according to the personal perception of the participant towards the main target group of East Belgium as a micro-destination.

After identification of the aforementioned characteristics for each persona, a pro forma 'customer journey' was created. The archetypical customer journey was designed based on modern understanding of value-chain creation for new tourism destinations (Partale and Partale, 2019, p. 14). According to the present methodological approach for the micro-destination, each persona follows the same customer journey (Fig. 6.2), but the touristic experiences are clearly different. The aim is to choose individually a touchpoint for each of the phases presented in Fig. 6.2 and to explain quickly why such touchpoints have been chosen for the persona.

This exercise was aimed at understanding the behaviour, reflections and thoughts of each persona and in turn understand where the focus regarding tourism product development and related online marketing needs to be. The most important results regarding the persona description and the customer journey are the following. The three experts decided that seniors, families and young groups were the most important target groups in their experience. Every buyer persona was very different from the other, which made the game very interesting and valuable. Interestingly, they chose the three main nationalities that had been identified in the market analysis as well. Regarding the customer journey, most of them mentioned that travel planning and booking were undertaken via the internet and booking platforms. The transport used was in two cases by car and in one case by train. Another interesting fact is that nobody chose a hotel as accommodation for their persona. They said that holiday accommodations and camping fitted better for their personas and were also typical for the region. Overall, the main activities were 'biking and hiking', and sportive activities, mostly in Worriken. For the target group, 'family and groups' it was also very important to have options in case of bad weather, as the weather in East Belgium is often unpredictable. They also mentioned regional products in restaurants. In every case, they were ready for a follow-up, no matter whether it was a word-of-mouth recommendation or a registration for newsletters, or a rating of the activity or accommodation online. Regarding sustainability, there was only one person who had the characteristics of actively being 'aware of sustainability and travel in a sustainable way'. The same persona opted to travel by train. However, passively, the other

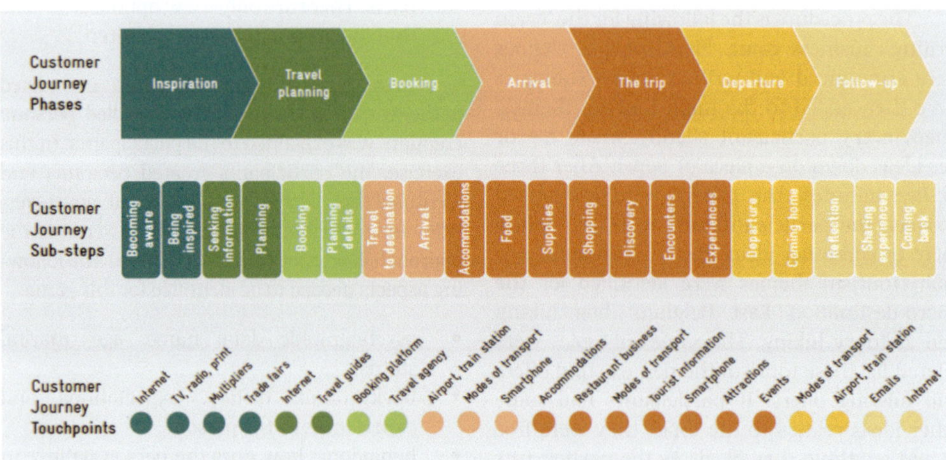


**Fig. 6.2.** Tourism-related customer journey for the micro-destination East Belgium. (Based on Partale and Partale, 2019.)

personas are sustainable as well as they mostly want to do activities in nature, without using a car. Furthermore, the persona who fits into the senior target group goes to the same camping site every year, which means that he acts sustainably from an economic perspective because he is very loyal to the camping site, and from a societal point of view because he interacts with locals.

Regarding the results of the two business games described above, the following framework (Partale and Traskevich, 2018) is used to discuss and to describe in detail the USP tourism themes for the micro-destination: travel motivations, target groups, memorable experiences, key resources and technologies, facilities, and touristic activities are identified and described for each USP theme. Once these key points are identified and described with the help of local experts and community representatives, the outcomes of the workshop can be applied to the strategy implementation and product design for the micro-destination East Belgium.

## 6.5 Product Design Considerations for the Micro-destination East Belgium

During the workshop, for each USP theme, several tourism themes were found: for the High Fens, seven different tourism segment themes were identified; for Worriken, four themes were created; and for the USP theme hiking and biking in combination with gastronomy, five new segment themes were identified. Target groups vary a lot. For particular segments, they are very niche-oriented, and for other segments, every target group can be attracted. On the results of the workshop, three segment themes of each USP that were elaborated for the micro-destination East Belgium were chosen to specify related development projects for these tourism themes. Such development projects are covered within a combination of different ideas from experts. The themes, which were chosen and discussed for elaborating the development projects are presented in Fig. 6.3.

All the nine development projects have been discussed and processed with the help of the results of the previously described workshop (Rauschen, 2020). Seven different segments, based on Partale and Traskevich (2018), were filled out. The motivation section is required in order to know what type of tourists would do the dedicated activity. The target group section defines the main target group (who does the activity when travelling to the micro-destination East Belgium). The experiences section elaborates the main reason to do the activity. People like to do activities to gain emotional, mental or physical benefits. The key resources and technologies provide the infrastructures required to participate in the tourism theme. The facilities are the local stakeholders of the micro-destination who contribute to the execution of the activity. Finally, the key activities represent the core of the elaborated tourism theme.

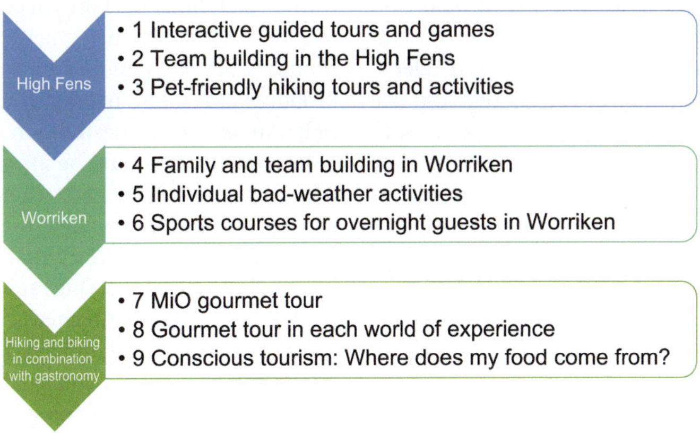


**Fig. 6.3.** Tourism development projects for each USP theme introduced for the micro-destination East Belgium. (Authors' illustration.)

**Motivation**

- People who are interested in regional products
- People who like to discover new products
- People who like to walk and are active

**Target group**

- Conscious tourists of all ages, segments and nationalities

**Experiences**

- Discovery of new tastes and flavours
- Discovery of landscape and culinary heritage

**Key resources and technologies**

- Regional producers
- Honeycomb system
- Vennbahn and Ravel paths

**Facilities**

- Regional restaurants and factories in the three worlds of experience: Brewery Bellevaux, Montenauer Ham, Belgium Peak Beer, Distillery Radermacher, Café Soiron and House Ternell,

**Key activities**

- MiO gourmet tour in the world of experience Eupen: biking
- MiO gourmet tour in the world of experience Malmedy: biking
- MiO gourmet tour in the world of experience Eifel: biking

**Fig. 6.4.** Description of the USP tourism segment theme: MiO gourmet tour. (Authors' illustration.)

While creating the nine development projects, the participants tried to elaborate on different themes that reach different target groups. These development projects try to reach not only one target, but to make the micro-destination interesting to a portfolio of demand groups. Projects one, three and four are oriented to broader target groups, but in particular they are attractive to families, seniors and 'best agers', along with young adults. Projects two and five are more for families or businesses who want to become stronger and more connected to each other in a familiar or working environment. Project seven is specifically aimed at all residents of Worriken. As Worriken is one of the USPs of the micro-destination, one project is dedicated to its residents only, to give them more input for their vacation. Lastly, projects seven, eight and nine are targeted at conscious travellers. They, thus, have various possibilities to contribute to the sustainability of the micro-destination. In this chapter, one segment theme is described in detail to provide a practical insight into the product design considerations elaborated in this study (Fig. 6.4).

As described previously, hiking and cycling in combination with gastronomy is a popular existing tourism product in the micro-destination East Belgium. However, there is no route which links MiO products and production facilities with existing tourist offerings. MiO products are an important component in the product value chain in terms of marketing East Belgium as a micro-destination. Such an added value component could be attractively compatible with the key product themes of culture, history and nature because it covers all three. It is, therefore, astonishing that there is no corresponding offer as yet. One reason for this could be that most production sites are not yet geared towards tourists and guided tours. However, not every MiO production facility may have a room where groups can come and have tastings during their walk.

Such an MiO gourmet tour can be best done by bike or e-bike, as visitor destinations are often a little too far apart for easy walking and it is difficult to cover the distances on foot in just one day. The MiO gourmet tours have been created for the three different worlds of experience: Eupener Land and Eupen, the High Fens and Malmedy and the Eifel and Sankt Vith. For each tour, three different stops have been planned to provide the guests with food and drink. They have the opportunity to get some explanations about the local product and receive a quick tour. This project appeals to all tourist senses. They see and smell nature, they taste and feel the local products and they hear about the history of the destination and the products. They benefit both emotionally and spiritually.

Other development projects are described within the same approach and the relevant details and creative solutions are achieved based on the previously described workshop. The experts and authors have come to the conclusion that elaborating on the USP tourism themes and related niche-oriented creative tourism products which base themselves precisely on local resources can be considered the key strategic approach in marketing a micro-destination as such.

## 6.6 Discussion

Within the aim of this study, creative strategic considerations for sustainable tourism development of the micro-destination East Belgium have been designed with intensive participation from local community representatives, tourism experts and policymakers, along with research on creative product design (Walder *et al.*, 2005; Partale and Traskevich, 2018; Partale and Partale, 2019, 2020; Prieto, 2020; Tomej and Xiang, 2020). Additionally, sustainable development of unique destinations and local communities, as well as the latest concepts on destination resilience, have been discussed (Dahles and Susilowati, 2015; Hall *et al.*, 2018; Ritchie and Jiang, 2019; Romão, 2020; Gössling *et al.*, 2021), along with the emphasis of this concept on local partnership and the intensive use of local tourism resources (Davoudi and Porter, 2012; Innerhofer *et al.*, 2018; Pechlaner *et al.*, 2019; Traskevich and Fontanari, 2021). Therefore, it can be concluded that the described methodological approach has proved to be useful in identifying new development projects for the micro-destination. At the same time, the empirical results also prove the unexploited tourist potential of the region itself.

The results lead to new findings on the topic of micro-destination. The proposed methodology can be applied to any kind of region but is particularly recommended for micro-destinations as they differ from neighbouring regions and have several USPs that can be investigated with the help of potential analysis, expert interviews and workshops with intensive involvement of local communities and regional stakeholders. The characteristics, attributes and necessities of each destination should be considered specifically whenever this methodology is applied to other kinds of tourism regions.

Through the unique and 'autonomous' culture of the region, the local community in East Belgium can be authentically defined by its diverse traditions that weld the separate villages together and bring people closer to each other. The people are very proud of their region. If the region is marketed as a micro-destination, locals will certainly find this motivating in terms of tourism development and try to convince others to come and explore their region. This emotional background and spiritual connection of local communities can be further considered the basis for the aforementioned innovative, uniqueness-oriented design of sustainable tourism products.

The entrepreneurial commitment is also quite high for micro-destinations. Various tourist companies are synergistically involved in making themselves and their region known. The Hotel Bütgenbacher Hof, for example, took part in a German TV show called *Mein Lokal, dein Lokal*. The hotel did very well in the evaluation, both in terms of the quality of the products and the service and design. This made it famous on German television. East Belgium was also discussed on the programme, highlighting the advantages and uniqueness of the region.

Through development projects, the relationship between locals and tourists can be further strengthened as they provide the relevant dialogical space that is attractive to both sides. Many of the projects require some communication between locals and tourists. The survival camp in the High Fens, which is part of another development project, only works if both locals and tourists get involved with each other. If both parties work against each other, the whole project will be negative and both parties will not be satisfied after the trip. Workshops based on farms also require mutual trust. The tourists are on the terrain of the locals, work with their machines and cows, and consume local products. If the tourists do not treat each other, and above all the locals, with respect, local people may get nervous or think negatively about the tourists and may become more closed to other tourists and projects. Such projects can be very time consuming for locals, but it is important to explain to them why good relations with tourists are so important. In this context, the results of the research confirm that the tools of internal marketing should be intensively employed towards the goals of flourishing of local uniqueness as well as overall sustainability and resilience.

If the local population is aware of sustainable tourism and its advantages regarding the economy, ecology and social life of the micro-destination, this will simplify communication between the local community and tourists, and facilitate the process of making the region more sustainable, attractive and resilient. Therefore, it is also important to encourage local tourism stakeholders to produce and work with MiO products as these products are of high quality and have a low carbon footprint. This is what makes the locals so hospitable, as they project their authentic behaviour onto how they deal with tourists. Due to the location of East Belgium, they also speak many languages, which often positively surprises tourists. They then feel all the more comfortable and welcomed in the micro-destination.

## 6.7 Conclusion

The current research has investigated the extent to which researchers have already dealt with the term 'micro-destination' and outlined the attributes assigned to the term. The research then examined to what extent the region of East Belgium can be called a micro-destination and which strategic considerations can be taken into account to guarantee sustainable tourism development there.

In the extensive market analysis of East Belgium, the attributes that make up the definition of a micro-destination were described in detail. These include, above all, the unique and diverging culture – for example, the language – and the homogeneity of locals, as well as the authentic history and nature of the region and, most notably, its particularity compared to the neighbouring regions. Existing USPs, current tourism projects, labels and international cooperation were presented as key factors in successful and sustainable tourism development at the level of a micro-destination. This resulted in an overall picture of the current tourism situation of the micro-destination East Belgium, where two USP tourism themes, namely the High Fens and hiking and biking in combination with gastronomy, could be identified. These USP themes were particularly important for pursuing the empirical part of this research, dedicated to elaborating new product design solutions for the micro-destination.

The presented methodological approach allowed us to find out how significant the tourism potential of East Belgium is and with which sustainable development projects this potential can be further exploited. For this purpose, two different business games were developed and an author workshop was applied. This workshop aimed to create different development projects and to identify motivations, target groups, experience, key technologies and resources, facilities, and key activities for each development project. By conducting a workshop with local experts and initiative community representatives, several strategies and development projects could be developed. Thus, the main objectives of the research – namely elaborating on the approaches to stimulate tourism demand, the improvement of marketing and brand awareness and better sustainable development of local communities – have been achieved within the proposed methodological approach to development strategy and product design in the micro-destination.

After the interview, a third USP (Worriken) and additional tourism products were identified, which set the starting point for the workshop. The results of the business games and overall critical discussion during the workshop showed that three main target groups could be identified: families, seniors and touristic groups. Furthermore, three different development projects could be created for each USP. In the case of the High Fens, related projects are interactive guided tours and games, team-building in the High Fens and pet-friendly hiking tours and activities. Regarding the USP segment Worriken, related development projects are individual bad-weather activities, family and team-building, and sports courses for overnight guests. In addition, for the segment hiking and biking in combination with gastronomy, an MiO gourmet tour in each world of experience, and conscious tourism ('Where does my food come from?') could be outlined. Each project was considered feasible and could be implemented if all stakeholders worked in partnership and demonstrated team spirit and patriotic unitedness.

The present contribution serves as the first cornerstone for further research in the field of micro-destinations. Such research should include the identification of other micro-destinations that fit the concept of strategic tourism development at the micro-level of regional governance. It should also reflect on the marketing aspect of micro-destinations and how they can be branded and promoted to become globally known.

A limitation of this research lies in further theoretical elaboration of the concept of tourism development at the level of a micro-destination, as neither the term nor the strategic concept has been sufficiently investigated, particularly from the perspective of conservation of local uniqueness through tourism development, sustainability and resilience. For further research, it would be interesting to find out how many regions fit the definition of a micro-destination that has been elaborated here. Then, it would be interesting to know if all these micro-destinations can be marketed as a whole, as suggested within the present methodological approach. Marketing many small regions under the same category and umbrella-brand of a micro-destination would probably make it easier to make them globally known in the tourism field – a factor initially indicated as the main limitation for the growth of international tourism in East Belgium.

Regarding the empirical section, it would be interesting to instigate further research into the synergies between tourists and locals and how this synergy could be improved, as it makes the stay more memorable and unique for tourists. At the same time, product design solutions should contribute to the preservation of the local uniqueness of micro-destinations, which require further practically oriented research towards sustainable tourism development.

## References


Andrades, L. and Dimanche, F. (2015) *Tourism in Russia: A Management Handbook*. Emerald Publishing Limited, Bingley.

Camillo, A.A., Minguzzi, A., Presenza, A. and Holt, S. (2015) Natural environmental sustainability and micro-tourism destinations: The case of southern Italy. In: Parsa, H.G. (ed.) *Sustainability, Social Responsibility, and Innovations in Tourism and Hospitality*. Apple Academic Press, Oakville, pp. 267–296.

Dahles, H. and Susilowati, T.P. (2015) Business resilience in times of growth and crisis. *Annals of Tourism Research* 51(2–4), 34–50. DOI: 10.1016/j.annals.2015.01.002.

Davoudi, S. and Porter, L. (2012) Resilience: a bridging concept or a dead end? *Planning Theory and Practice* 13(2), 299–333.

Gössling, S., Scott, D. and Hall, C.M. (2021) Pandemics, tourism and global change: A rapid assessment of COVID-19. *Journal of Sustainable Tourism* 29(1), 1–20. DOI: 10.1080/09669582.2020.1758708.

Hall, C.M., Prayag, G. and Amore, A. (eds) (2018) *Tourism and resilience: Individual, organisation and destination perspectives*. Channel View Publications. Series: Tourism Essentials: 5, Bristol.

Hernández-Martín, R., Simancas-Cruz, M.R., González-Yanes, J.A., Rodríguez-Rodríguez, Y., García-Cruz, J.I. *et al*. (2016) Identifying micro-destinations and providing statistical information: a pilot study in the Canary Islands. *Current Issues in Tourism* 19(8), 771–790. DOI: 10.1080/13683500.2014.916657.

Heukemes, N. (2009) *Regional Development Concept of the German-speaking Community*. Kliemo AG, Eupen.

Innerhofer, E., Fontanari, M. and Pechlaner, H. (2018) *Destination Resilience: Challenges and Opportunities for Destination Management and Governance*. Routledge (Taylor & Francis Group), Oxford.

Ostbelgien (2018) Regional products made in East Belgium. Available at: https://www.ostbelgien.eu/en/what-to-do/discover/regional-products (accessed 7 May 2020).

Ostbelgien (2019) *Holiday catalogue 2020*. Kliemo AG, Sankt Vith.

Ostbelgienlive (n.d.) East Belgium Live – Location and scope. Available at: https://www.ostbelgienlive.be/desktopdefault.aspx/tabid-84/187_read-450/ (accessed 28 April 2020).

Ostbelgientourismus (2019) Quality initiative. Available at: http://www.ostbelgientourismus.be/desktopdefault.aspx/tabid-5121/9030_read-59184/ (accessed 10 June 2020).

Papatheodorou, A. (2006) Microfoundations of tourism choice. In: Dwyer, L. and Forsyth, P. (eds) *International handbook on the economics of tourism*. Edward Elgar, Cheltenham, pp. 73–88.

Partale, A. and Partale, K. (2019) *Destination management in developing and emerging countries: Handbook and guidelines for building sustainable destination management organizations*. Deutsche Gesellschaft für Internationale Zusammenarbeit (GIZ), Bonn.

Partale, A. and Partale, K. (2020) *Die Wertschöpfungskette Tourismus: Analyse und Anwendungsansätze für Projekte der Entwicklungszusammenarbeit* [The tourism value chain: Analysis and application approaches for development cooperation projects]. Deutsche Gesellschaft für Internationale Zusammenarbeit (GIZ), Bonn.

Partale, K. and Traskevich, A. (2018) Methodological approach to the tourism development strategy and its application to Naroch destination. *Belarusian Economic Journal* 20(1), 131–146.

Pechlaner, H., Zacher, D., Eckert, C. and Petersik, L. (2019) Joint responsibility and understanding of resilience from a DMO perspective – an analysis of different situations in Bavarian tourism destinations. *International Journal of Tourism Cities* 5(2), 146–168. DOI: 10.1108/IJTC-12-2017-0093.

Prieto, M. (2020) *Design Thinking for Tourism Product Development: Application of Design Thinking Techniques to Achieve Sustainable and Memorable Tourism Experiences*. Sciencia Scripts, Beau Bassin, Mauritius.

Rauschen, L. (2020) Strategic considerations for sustainable tourism development of the micro-destination East Belgium. MSc thesis. Hochschule Fresenius University of Applied Sciences, Cologne, Germany.

Ritchie, B.W. and Jiang, Y. (2019) A review of research on tourism risk, crisis and disaster management: Launching the annals of tourism research curated collection on tourism risk, crisis and disaster management. *Annals of Tourism Research* 79(1), 102812. DOI: 10.1016/j.annals.2019.102812.

Romão, J. (2020) Tourism, smart specialisation, growth, and resilience. *Annals of Tourism Research* 84, 102995. DOI: 10.1016/j.annals.2020.102995.

Schmitz, C. (2011) VAO now called Tourism Agency East Belgium. Available at: https://www.grenzecho.net/art/zz/hier-und-heute/vao-heisst-jetzt-tourismusagentur-ostbelgien (accessed 10 June 2020).

Schoop, S. (2019) *Digital Innovation*. slide 233-238, HS Fresenius Cologne.

Segota, T. (2018) (G)A(i)ming at the Throne: Social media and the use of visitor-generated content in destination marketing. In: Ziakas, V. and Lundberg, C. (eds) *The Routledge Handbook of Popular Culture and Tourism*. Routledge, Milton, pp. 427–438.

Stancioiu, A., Teodorescu, N., Pargaru, I., Vladoi, A.D. and Baltescu, C. (2011) The image of the tourism destination – a supporting element in the development of the regional tourism brand study case: Muntenia. *Theoretical and Applied Economics* 2(555), 139–152.

TAO Beratungs- und Management GMBH (2013) *Tourism Development East Belgium 2025 Mission Statement*. Regierung der Deutschsprachigen Gemeinschaft, Villach.

Tomej, K. and Xiang, Z. (2020) Affordances for tourism service design. *Annals of Tourism Research* 85, 103029. DOI: 10.1016/j.annals.2020.103029.

Traskevich, A. and Fontanari, M. (2021) Tourism potentials in post-COVID-19: The concept of destination resilience for advanced sustainable management in tourism. *Tourism Planning & Development* 82, 1–25. DOI: 10.1080/21568316.2021.1894599.

Walder, B., Weiermair, K. and Sancho Pérez, A. (2005) *Innovation and Product Development in Tourism*. Erich Schmidt Verlag, Berlin.

# 7 Perceived Impacts of Urban Tourism on Host Communities: Comparing Milan and Porto

**Valeria Fusè[1]* and Elisabeth Kastenholz[2]**
*[1]Università degli Studi di Bergamo (ITA), Bergamo, Italy; [2]Research Unit GOVCOPP (Governança, Competitividade e Políticas Públicas), Universidade de Aveiro, Campus Universitário de Santiago, Aveiro, Portugal*

## 7.1 Introduction

Urban tourism boomed in many European destinations prior to the Covid-19 pandemic crisis, with large numbers of travellers engaging in short break holidays and weekend trips during the year, probably fostered through low-cost airline connections and cheap accommodation opportunities available on competitive booking and sharing platforms (Freytag, 2009; Ioannides *et al.*, 2019). The impacts of these increasing tourist flows on many cities have added to their complexity and dynamism, with both positive and negative effects visible and discussed in the tourism literature (Sharpley, 2006; Hayllar *et al.*, 2008a, b; Ashworth and Page, 2011; Namberger *et al.*, 2019). Many cities need to reinvent themselves in times of global competition and to redefine their main functions, with particularly urban, frequently cultural, tourism standing out as strategies of urban development. This chapter discusses the challenge of integrating tourism within a perspective of sustainable urban development, presenting a study of the local communities' view of two important European cities, Porto and Milan, which have placed tourism and culture at the heart of their development.

For both travellers and local residents, tourism represents a relevant shared experience, leaving impacts on all involved, but the accumulated effects of tourist flows are particularly significant for cities' inhabitants. They need to cope with increased demand on their public spaces and services and renewed or redefined cultural identities, while benefiting from stimulated economies and a greater variety and complexity of social life. Negative economic, social and environmental impacts cannot, however, be ignored. Although research on residents' attitudes towards tourism is nothing new, most studies use quantitative methods and only a single case analysis (Sharpley, 2014), while qualitative and comparative studies are relatively rare. The present study addresses these gaps and advances an understanding of the impact of tourism on residents of one of the typically most attractive sites in European cities – their historical centres – finding differentiated views through a mixed qualitative–quantitative approach, interviewing residents of city centres in Porto and Milan. These views are additionally contextualized in terms of the phase of tourism development and

*Corresponding author: valeriafuse099@gmail.com

 *Tourism Planning and Development in Western Europe* (eds K. Andriotis *et al.*)
DOI: 10.1079/9781800620797.0007

the type of urban planning, permitting comparisons and a debate on reasons for differences, while also identifying similar patterns of impact perception and attitudes.

The next section of this chapter introduces urban tourism and presents a brief examination of urban cultural policies in the development of cities as tourist destinations. Then, drawing on the existing literature on the role of communities and the importance of impact perception for sustainable urban development, this study advances an understanding of the host–guest relationships in urban tourism. The following section presents the case studies analysed, including a brief presentation of the background of each city destination and the research methodology. The results appear in the next section, describing the samples and respondents' reported interaction with tourists and their perceptions and attitudes towards tourism. To conclude, the empirical evidence confirms previous findings regarding the important role of social interaction in the urban tourism experience and demonstrates that residents subjectively interpret these interactions according to tourists' behaviour and their own interaction with them, while the particular context of urban development, and of the integration of tourism within it, plays a central role in understanding differences in locals' perception of tourism impacts.

## 7.2 Urban Tourism

The United Nations World Tourism Organization (UNWTO) defines urban tourism as 'trips taken by travellers to cities or places of high population density. The duration of these trips is usually short (one to three days), making them closely linked to the short-breaks market' (UNWTO, 2002; Tourism 2020 Vision). These trips have recently gained increasing academic attention. However, urban tourism's multi-functional nature has contributed to a lack of a consensual definition, despite its consolidation as an established topic in the growing tourism research. Being inherently different from other forms of tourism (Ashworth and Page, 2011), urban tourism entails its location in an urban spatial context, with which a close connection in the consumption of tourist products is established. Tourism constitutes a key determinant in shaping the spatial, social and economic attributes of the destination and it appears as a vehicle for economic development with a variety of attractions and facilities to sell.

Urban areas are characterized by a perception that their landscape is a web in which many individuals and groups have social, cultural, political and economic interactions and relationships. On the one hand, urban environment has a direct impact on social wellbeing, affecting citizens' behaviour and culture, and on the other hand, residents improve their living conditions by actively co-creating an environment of social and cultural dynamism and appeal. The UNWTO (2012) affirms that 'the growth of the city and the ultimate goal of urban planning should be the logical response to the needs of its residents in order to increase their quality of life: wellbeing, environmental quality and cultural identity'. City managers and planners thus have to respond to the challenges of the spatial configuration of the city in order to ensure the most favourable conditions of life, benefiting both residents and visitors alike. Accordingly, in recent years, tourism has been recognized as an important element in urban policies, representing both new challenges and opportunities, and calling for a careful management of its flows, concentration and effects in urban environments.

The United Nations (2019) estimates that 55% of the world's population lives in urban areas and is expected to represent 68% by 2050. As cities are rapidly growing, managing the increasing numbers of those using these spaces is of primary importance to ensure sustainable urban development. In addition, with the proliferation of online booking, city tourism services are becoming easily accessible and urban destinations are constantly changing to adapt to the growing tourism demand by providing a wide offer of cultural, natural, architectural and technological products and experiences for leisure and business. Considering that urban tourism can bring new patterns of space consumption, as well as transformations to the urban landscape and living context, there is a need to study more closely the city and its development in respect of such changes, with the aim of providing well-adapted urban policies.

### 7.2.1 Redefining the city

During the last decades of the 20th century, significant changes occurred in cities' development. Many authors have highlighted a correlation between urbanization and economic development. Accordingly, the creation of global networks facilitates the capital investment flows that simultaneously link and separate cities by fluctuating in and out of the urban economy and spatial structure (Stevenson, 2003). The changes in economic power and capital, being progressively more mobile, have affected the role of cities: from being centres of production of goods and services, to rather nodal connectors of wider international networks. Predominantly, cities are shifting from being sites and spaces of mass production towards reinventing themselves as centres of consumption (Law, 1996, p. 599). As the forces of globalization place cities into direct competition with one another, cities are today required to deliver thriving economies, greater quality of life, political stability, business friendliness and a reduced environmental impact in order to be competitive, not only on a regional or national scale, but globally (Peris-Ortiz *et al.*, 2017). In addition, cities have realized that they have to attract an increasingly mobile consuming public in order to support their local economies (Richards, 2000). Therefore, feeling the effects of globalization, cities have started to take strategic action to enhance their performance and competitiveness at a world level. To differentiate themselves from others, more and more cities enact regeneration strategies, whose driver of growth is the cultural economy, including the whole city, its heritage, material development and local initiatives flowing into a wider cultural policy fostering a sense of place and a unique destination image. In this context, cities are increasingly using urban cultural policies as an element of city governance, offering both visitors and residents a unique place to live, work, visit and invest, especially in Western Europe (García, 2004). These strategies are implemented by adopting cultural projects for revitalizing city spaces, as well as reusing and rebuilding programmes for those spaces that have lost their previous function (Galdini, 2019).

Central to such strategies are local diversity and territorial identities, which assume a pivotal role in the distinctiveness of the city at the global level (Ray, 1998). The importance of 'localism' emerges as one of the effects of globalization, in which global competition among different cultures and urban destinations may intensify the preservation of local identities, even within apparently highly globalized cities like Singapore (Chang *et al.*, 2004). To this end, urban cultural planning has focused on the increasing salience of local specificities, making them attractive at a global level by stimulating creative use of such specificities, for example, among other things, integrating authenticity in designing touristic routes or offering local experiences (Richards, 2014). Nevertheless, a major problem with these regeneration strategies is that urban governance aims at using culture, or specifically local culture, to produce economic growth by addressing the mobile consuming public and foreign investors, rather than local communities and cultural agents. As a consequence, the role of cultural policy appears to include the local, regional and national cultural identities to generate cultural diversity at the global level, commoditizing them as simply knowledge economy assets. Therefore, many researchers have pointed out that the main contribution of cultural policies to urban regeneration is probably the construction of place image, leading to increased tourist flows and corresponding new service functions. However, the impacts of this policy on wealth and employment creation in the local cultural industry, and on real cultural development at the local level, are often questionable.

## 7.3 The Role of Local Communities in Sustainable Urban Development

In urban tourism, the tourist experience itself and the way in which tourists travel to and move within an urban destination mirrors their motivations, preferences and cultural perspectives. Specifically, urban tourism experiences evolve through visitors' interactions with local people, as users of the same space, sharing the same amenities, services and infrastructures, which additionally represent part of local culture and help in the perception of those experiences as more authentic. On the other hand, the host community's subjective perceptions of tourism are important because these perceptions affect the behaviour of residents toward tourists (Andriotis and Vaughan, 2003). Thus, the local community's goodwill and cooperation are critical for the destination's success (Stylidis *et al.*, 2014; Bimonte and Punzo, 2016), which requires the involvement of all relevant stakeholders for achieving sustainable development. In this context, guaranteeing the quality of life of the host community becomes of primary importance in sustainable tourism planning.

The academic literature studies host–guest relationships by mainly focusing on the residents' side and usually together with their perceptions of the impacts of tourism. It has emerged that host–guest interactions vary according to the specific conditions of each urban setting, as do the real and perceived tourism impacts, which of course influence the residents' quality of life and their attitudes towards tourism and tourists. Among the conceptual frameworks, Doxey's (1975) 'Irridex' and Butler's (1980) Tourist Area Life Cycle are the most famous models developed to theorize residents' reactions towards tourism. While the former claims that locals' attitudes evolve along four stages of acceptance of tourism from euphoria to antagonism, the latter states that the evolution of tourist destinations follows six steps of development from exploration to decline or rejuvenation. These models suggest that the presence of tourists is generally accepted by local residents, as long as their negative impacts remain below a critical level of tolerance. In addition, the nature and form of host–guest interactions and consequent attitudes towards tourism vary in accordance with specific place context (de Kadt, 1979), the frequency of interaction and with residents' involvement with the tourism industry (Krippendorf, 1987; Sharpley, 2014), influencing both the tourist experience and the community's perceptions of tourists and tourism impacts. Indeed, positive interaction and relationships between locals and tourists are of crucial relevance for the satisfaction of both actors involved (Eusébio and Carneiro, 2012a, b; Kastenholz *et al.*, 2013).

As far as impact perception is concerned, conceptually, positive and negative tourism impacts are usually studied according to the Triple Bottom Line (TBL) approach, covering three dimensions: the economic, the sociocultural and the environmental impact (Ngowi and Jani, 2018). These consequences of tourism influence local communities, generating benefits and/or triggering damage and costs. Additionally, Social Exchange Theory (SET) is frequently considered for explaining impact perceptions. It recognizes the host community as being a heterogeneous group with different attitudes according to each individual's perception of the costs and benefits of tourism, and also from a personal perspective of being involved and personally affected by the industry (Nunkoo *et al.*, 2013). However, there is no strong consensus regarding the factors that affect residents' attitudes, suggesting their variation according to the specific conditions of each environment. Accordingly, urban policymakers and planners need to ensure sustainable, community-integrating urban development, considering the city's specific cultural, social and economic structure, and integrating tourism regeneration strategies within sustainable development strategies for host communities, ensuring the preservation of the environment, cultural heritage, social wellbeing and income generation (Wise, 2016). This goal increasingly calls for a control of tourist flows, avoiding overtourism.

## 7.4 The Cases of Milan and Porto

Within this framework, this empirical research is concerned with the particular features of two European cities, namely Milan and Porto (Figs 7.1 and 7.2), which represent two interesting case studies among other urban destinations, showing both similarities and differences in their context and tourism development. Both cities, given their cultural significance, present themselves as adequate frameworks for the study of tourism impacts in urban spatial contexts, regarding specifically host–guest interaction and residents' perceptions of urban development. As a matter of fact, both cities have recently been recognized as 'global cities' by the Globalization and World Cities Research Network (GaWC, 2010, 2018), considering their connectivity in the world city network. Both Milan and

**Fig. 7.1.** The City of Milan with its landmark at the centre. (Photo by Ouael Ben Salah on Unsplash.)

**Fig. 7.2.** Porto Historical Centre as UNESCO's World Heritage Site. (Photo by Miguel A. Ferreira from Pexels.)

Porto are the second major urban areas within their respective country, and represent an international reference in urban tourism, being well-known destinations for national, European and global tourists (European Best Destinations, 2017).

Certainly the two cities differ, firstly, in terms of natural morphology. Milan is located in the middle of a plain surrounded by the foothills of the Alps, while Porto rises over granite hills and reaches the Atlantic Ocean. Furthermore, in terms of the dimensions of the urban areas and number of population, the city of Porto counts 222,252 inhabitants (Câmara Municipal do Porto, 2011) while the city of Milan has a population of 1,366,180 (ISTAT, 2018). However, both cities have shared a spatial transformation process, evolving from industry-based economies to mainly service-based economies, placing culture at the centre of their urban development. Urban cultural planning has increased their attractiveness, offering both visitors and residents a unique place to live, work, visit and invest. In particular, the city of Milan has transformed into an international creative hub, with the creation of the new cultural and creative districts such as Isola and Tortona and the use of mega-events as a means to enhance its international profile, such as the 2015 International Exposition and the 2026 Winter Olympic Games, attracting more than 3 million foreign tourists (lidisSAS Visual Analytics Viewer, 2015) for both business and leisure reasons.

On the other hand, Porto was an important industrial city until 1996, when its historical centre was recognized as a UNESCO World Heritage Site, which required several interventions in the area to protect its historical value. The city's nomination as 'European Capital of Culture' in 2001, further favoured urban cultural policies, which today attract 1.6 million tourists (INE, 2017). Along with the growing market interest in urban tourism, the two cities have just recently gained international recognition and affirmed themselves as important tourist destinations, with Porto being elected best European destination in 2012, 2014 and 2017 (Wine Tourism Portugal, 2017). As a result of the recent tourism boom in this Portuguese city, Airbnb offerings have increased rapidly, making the city stand out as one of the ten European cities with the highest numbers of Airbnb listings per 1000 inhabitants (Adamiak, 2018).

### 7.4.1 Methodology

The study aimed to examine the nature and form of host–guest relationships and the residents' perceptions regarding the impact of tourism in two specific cities: Porto and Milan. On the one hand, the research sought to identify the residents' relationship with tourists according to the frequency, mode and place of interaction, as well as the quality of their contact, as individually perceived. Specifically, the study focused on the individual experiences and personal evaluations of such interactions. On the other hand, the impacts of tourism in these urban settings, particularly the historic city centres, were also investigated, taking into account the three main sustainability dimensions – economic, social and environmental – and considering both positive and negative aspects. A mixed methods research approach was adopted to answer the research questions and to understand many different aspects of the mentioned urban tourism phenomena, from the residents' perspective, while also permitting diverse types of analysis. Both a questionnaire and a semi-structured interview were personally administered to the residents of both cities to collect both qualitative and quantitative data. This was done by applying a questionnaire with open-ended questions personally, with closed-ended questions either filled in by the respondent or by the interviewer (as preferred by the interviewee). The open-ended questions were filled in by the interviewer, who also asked for permission to audio-record the responses to work on details later. Most respondents accepted audio-recording.

As far as the cities of Porto and Milan are concerned, official statistics on the resident population helped define an approximate quota sample for the historical centre population, taking into account age, gender and parish of residence. The main author of the chapter conducted interviews personally with 30 residents of Porto during one month of a research stay, and with residents of

Milan for almost three weeks. A combined quota and a 'snowball' sampling strategy were applied, the final approach being used to fulfil the specified quota criteria. The interviewer being an Italian native speaker, not fluent in Portuguese, but having some Spanish, while the literature base for the questionnaire development was English, the questionnaire/interview guide was first elaborated in English. Accordingly, some basic knowledge of English was considered a sampling criterion for the Porto residents, although Spanish and Portuguese were also accepted in special circumstances, particularly when the resident was able to answer the translated questions slowly or if the interviewer was able to communicate with the Portuguese resident with the help of any accompanying person. Later, for the application of the research instrument in Italy, the questionnaire was translated into Italian.

The questionnaire includes the following three sections:

1. Type of social interaction with tourists.
2. Residents' attitudes toward the three dimensions of tourism impacts.
3. Sociodemographic variables.

Suitable questions/items for assessing these categories were extracted from a literature review, including the following aspects.

1. **Type of social interaction with tourists**, in terms of (a) frequency of interaction; (b) mode of interaction; (c) personal attitude of interaction; (d) place of interaction; (e) quality of interaction.
2. **Residents' attitudes towards the three dimensions of tourism impacts**
   **2.1 Economic impacts**: (a) prices of goods and services; (b) employment opportunities; (c) housing prices; (d) local taxes.
   **2.2 Social impacts**: (a) cultural activities; (b) public services and safety; (c) crowding; (d) traditions and values; (e) local identity.
   **2.3 Environmental impacts**: (a) city's image in terms of attractiveness; (b) preservation of historical buildings; (c) number of green areas; (d) maintenance of green areas; (e) authenticity of cultural heritage; (f) quality of roads, bridges, infrastructures; (g) quality of air, water; (h) noise and sound pollution.
3. **Sociodemographic variables**: (a) length of living in the area; (b) job's dependence on tourism; (c) gender; (d) age; (e) level of education; (f) monthly household's average income.

The interaction type and personal attitudes were assessed using a 5-point scale from 1 'totally disagree' to 5 'totally agree' for 20 items associated with the experiences and feelings of social relations with tourists (Eusébio and Carneiro, 2012a, b; Kastenholz *et al.*, 2013). A 5-point scale from 1 'very negative experience' to 5 'very positive experience' was used to assess the overall quality of the interaction with tourists. The perceptions of the three tourism impact dimensions were measured using a 5-point scale from 1 'extremely decreased/worsened' to 5 'extremely increased/improved' (in which 3 was 'no effect') for 17 items reflecting the perceived changes in urban settings through tourism development (Andereck *et al.*, 2005; Eusébio and Carneiro, 2012b).

As noted, the questionnaire was designed in English and then translated into Italian, with 30 responses collected in both languages. For Spanish and Portuguese speakers, a simultaneous translation was provided. As also noted, after receiving respondents' agreement, interview responses were audio-recorded. Interviewees tended to give a short reason for their choices, specifying their opinions and experiences, in particular with regard to the open-ended questions. The latter were transcribed, subjected to content analysis and then categorized in an attempt to recognize the main issues of each respondent's discourse and to ease the comparison with the quantitative data and between the study cases. All quantitative and categorized data were introduced into an SPSS (version 24) database. The statistical data analyses comprised descriptive and non-parametric analyses to investigate associations within the data.

## 7.5 Results

A total of 60 responses were collected, 30 in the city of Porto and 30 in the city of Milan. The sample aimed at being approximately representative of both populations, taking into account age ranges and gender and also with the concern of leading to theoretical saturation regarding topics being discussed. A total of 32 female and 28 male individuals were interviewed. The largest group of residents was between the ages of 35 and 64 years (43% of the sample) while 32% were under 34 years old and only 25% above 65 years. While Italian was used for all the interviews in Milan, in Porto, most of the conversations were held in English, one third in Portuguese and 10% in Spanish.

The sample had more respondents with secondary education, corresponding to 37% of the total. In terms of income, about one third of the participants reported a family monthly net income between €2001 and €4000, with Porto residents showing lower levels of income. Most of those interviewed were long-term residents and reported they had been living in the central district of each city for more than 25 years (48% of the sample).

With respect to job dependence on tourism, the majority of respondents reported that their job or activity does not imply any contact with tourists (65%), with 31% reporting direct or indirect contact with tourists. Table 7.1 presents these results.

Work-related contacts were found more often for Porto residents (50% reporting direct or indirect contact), explaining more interaction opportunities with tourists in the Portuguese city. Host–guest interactions seem to be related to the residents' job dependence on tourism, which is not surprising. The number of direct and indirect contacts is higher than expected in daily situations, especially in Porto.

Focusing on interactions next, the majority of the interactions generally took place in public spaces and on public transportation (65%), in open-spaces and on the street (73%), as well as in other facilities such as cafés, restaurants, shopping malls, cinemas, etc. (60%) (Table 7.2).

Looking at the form and nature of the host interaction with tourists, one may observe that respondents have generally exchanged information about the place while interacting (3.8 in a scale from 1 'totally disagree' to 5 'totally agree'). While interacting with tourists, residents stated that there have been moments of entertainment (3.8) and they have had the opportunity of getting to know other cultures (3.7) (Table 7.3).

Host–guest interaction is more frequent when (a) seeking the opportunity of getting to know other cultures; (b) exchanging information

**Table 7.1.** Sociodemographic profile and travel context (highest values in bold).

| Demographic | Total $n = 60$ |
|---|---|
| **Gender** | |
| Female | 32 |
| Male | 28 |
| **Age range (years)** | |
| <19 | 10 |
| 20–34 | 9 |
| 35–49 | 12 |
| 50–64 | **14** |
| 65–79 | 10 |
| >80 | 5 |
| **Language of interview** | |
| English | 18 |
| Portuguese | 9 |
| Spanish | 3 |
| Italian | **30** |
| **Highest level of education** | |
| Primary school | 12 |
| Secondary school | **22** |
| Bachelor's degree | 12 |
| Master's or doctoral degree | 14 |
| **Average family monthly net income (€)** | |
| <500 | 4 |
| 501–1000 | 8 |
| 1001–2000 | 10 |
| 2001–4000 | **17** |
| 4001–6000 | 7 |
| > 6000 | 3 |
| **Length of living in the neighbourhood (years)** | |
| <5 | 8 |
| 5–10 | 1 |
| 10–15 | 7 |
| 15–20 | 7 |
| 20–25 | 8 |
| >25 | **29** |
| **Job dependence on tourism** | |
| Work in tourism field | 2 |
| Direct contact | 11 |
| Indirect contact | 8 |
| No contact | **39** |

**Table 7.2.** Interaction-related variables (highest values in bold).

| Variable | Total $n = 60$ |
|---|---|
| **Frequency of the interaction** | |
| Daily | 13 |
| Few times a week | 5 |
| Once a week | 8 |
| Few times a month | **15** |
| Once a month | 5 |
| Less once a month | 14 |
| **Place of interaction** | |
| **Public spaces and transportation** (such as trains and bus stations, airports, etc.) | |
| No | 21 |
| Yes | **39** |
| **Private spaces** (such as home, car, Airbnb, accommodation, etc.) | |
| No | **47** |
| Yes | 13 |
| **Street-interaction and open spaces** (such as green areas, parks, squares, etc.) | |
| No | 16 |
| Yes | **44** |
| **Other facilities** (such as restaurants, cafés, libraries, cinemas, stores, etc.) | |
| No | 24 |
| Yes | **36** |
| **Other – study hall** | |
| No | 59 |
| Yes | 1 |
| **Other – telephone** | |
| No | 59 |
| Yes | 1 |
| **Other – club** | |
| No | 59 |
| Yes | 1 |

about the place with tourists; and (c) participating in events or parties with tourists, as observed through ANOVA testing of differences between interaction type and frequency of interaction (with *p*-values of 0.027, 0.032 and 0.030 respectively).

Respondents evaluated their experiences with tourists as mostly positive (4.3, on a scale from 1 'very negative experience' to 5 'very positive experience'). The experience has also been emotionally rewarding, since people mostly 'enjoyed interacting with tourists' (4.3 on a scale from 1 'totally disagree' to 5 'totally agree'). In addition, residents have a very favourable perception of and emotional reaction to experiencing the interaction with tourists, as they reported a high level of agreement in terms of feeling stimulated (4.1), curious (3.9), and enthusiastic (3.9). Another important feeling resulting from interaction with tourists is increased emotional attachment to their city (3.6).

As far as residents' perceptions towards the impacts of tourism on their urban environment are concerned, results do not reveal any perception of excessive change regarding any impact dimension. However, the findings reveal a global awareness of the cities' improved image in terms of attractiveness in both cases (4.4 on a scale from 1 'extremely decreased/worsened' to 5 'extremely increased/improved') (Table 7.4).

By considering each dimension of the impacts of tourism, every economic item has mostly increased according to respondents' perceptions, particularly prices of housing (4.2), goods and services (3.8), local taxes (3.8) and employment opportunities (3.8), the only positive impact item in the list. Regarding the sociocultural dimension, respondents stated that there has been a rise in crowding (4.1) as a negative impact. Increases in cultural activities (4.0) and a small improvement in public services and safety (3.5) were deemed positive consequences of tourism.

In relation to the environmental impacts of tourism, besides the enhancement of the cities' overall attractive image, respondents highlighted a better preservation of historical buildings (3.7). Apart from this, a slight increase in the number of and improvement or maintenance of green areas as well as an enhancement of authenticity of cultural heritage was reported as a positive effect, while a moderate increase in noise and sound pollution was regarded as a negative impact. It is interesting to observe that respondents perceived a decrease in the quality of air and water associated with tourism development, as shown in Table 7.4.

**Table 7.3.** Interaction statements (highest values in bold).

| Statement | Mean | Std deviation |
|---|---|---|
| *Mode of interaction* | | |
| I have exchanged information about the place with tourists | **3.8** | 1.176 |
| There have been moments of entertainment while interacting with tourists | **3.8** | 1.242 |
| I have been taking photos with tourists | 2.8 | 1.531 |
| I have had the opportunity of getting to know other cultures through interaction with tourists | **3.7** | 1.374 |
| I have invited tourists to stay at my home | 1.8 | 1.482 |
| I have participated in events or parties with tourists | 2.6 | 1.624 |
| I have engaged in recreational activities with tourists | 2.6 | 1.630 |
| I have visited monuments or the historical centre together with tourists | 2.3 | 1.497 |
| I have participated in cultural activities with tourists | 2.3 | 1.485 |
| I have shared meals with tourists | 2.2 | 1.474 |
| I avoid interaction with tourists | 1.2 | 0.856 |
| *Personal attitude towards interaction* | | |
| The interaction with tourists has stimulated my curiosity | **3.9** | 1.160 |
| The interaction with tourists made me feel enthusiastic | **3.9** | 1.263 |
| The interaction with tourists made me feel surprised | 2.9 | 1.329 |
| After the interaction, I have felt more attached to my city | 3.6 | 1.186 |
| I enjoyed interacting with tourists | **4.3** | 0.917 |
| The interaction with tourists made me feel stimulated | **4.1** | 1.136 |
| Interacting with tourists made me feel angry and irritated | 1.2 | 0.493 |
| The interaction made me feel bored | 1.1 | 0.547 |
| The interaction with tourists made me feel lonely | 1.1 | 0.468 |
| The interaction with tourists made me feel sad | 1.1 | 0.362 |
| *Quality of the interaction* | | |
| Experience from contact with tourists | **4.3** | 0.725 |

Note: Mean values refer to mean ratings on 5-point Likert scales (ranging from 1, do not agree, to 5, totally agree).

The study did not demonstrate any association between these impact perceptions and the length of living in the neighbourhood, thus not confirming any association between longer living periods and more intense impact perceptions, as initially assumed. The qualitative data in this mixed-method data collection help additionally interpret some of the survey results, with three main themes identified: the experience of interaction; the changes in the urban setting; and the effects of tourism. The interview data are discussed next.

### 7.5.1 The City of Porto

Residents interviewed in Porto described their interactions with tourists mostly positively. They generally declared to have good relationships with tourists, evaluating tourists as friendly and their behaviour as nice. Few of them revealed an unpleasant experience of encounters with English speakers because of their concern in having a conversation in a foreign language. The interviews portray Porto as a very cosmopolitan city, since every respondent answered that he/

**Table 7.4.** Perception of tourism impacts (highest values in bold).

| Item | Mean $n = 60$ | Std. deviation |
|---|---|---|
| *Economic impacts* | | |
| Housing prices | **4.2** | 0.913 |
| Prices of goods and services | **3.8** | 0.867 |
| Employment opportunities | **3.8** | 0.909 |
| Local taxes | **3.8** | 0.874 |
| *Sociocultural impacts* | | |
| Cultural activities | **4.0** | 0.622 |
| Crowding | **4.1** | 0.915 |
| Traditions and values | 3.2 | 0.890 |
| Public services and safety | 3.5 | 0.892 |
| Local identity | 2.9 | 1.048 |
| *Environmental impacts* | | |
| City's image in terms of attractiveness | **4.4** | 0.588 |
| Preservation of historical buildings | **3.7** | 0.841 |
| Number of green areas | 3.6 | 0.715 |
| Maintenance of green areas | 3.6 | 0.810 |
| Authenticity of cultural heritage | 3.5 | 0.833 |
| Noise and sound pollution | 3.5 | 0.910 |
| Quality of roads, bridges, infrastructure | 3.3 | 0.799 |
| Quality of air, water | 2.6 | 0.696 |

Note: mean values refer to mean ratings on 5-point Likert scales (ranging from 1, do not agree, to 5, totally agree).

she had interacted with tourists from all over the world, i.e. French, Brazilian, British, Canadian, Swiss, Iranian travellers, etc. The most positive interactions occurred mainly with Spanish visitors, probably thanks to the similarity of language and proximity of culture (Kastenholz, 2010) or because they may represent the largest share of the foreign tourist market in Porto.

Considering the changes perceived in the old town and the riverside, also called 'Ribeira', respondents recognized a positive development thanks to the regeneration programmes implemented by the municipality. They described a transformation from an old neighbourhood to a historical centre, where the old has been turned into added value through tourism development, and believed the area to be 'overall more attractive, more touristic and more lively [...]' (woman, 50–64 years). While this was true for many of those interviewed, the younger ones in particular appreciated this vitality and felt stimulated to walk around, even during the night and on weekends, while others felt oppressed by the number of tourists, street congestion and crowding. There seems to be a divide in interpretation here, according to preference for more or less crowded (or lively) spaces.

Nevertheless, the respondents revealed unanimity regarding the economic benefits that tourism has brought to Porto, which must be understood in the light of the relatively recent and impactful financial crisis that Portugal has gone through, mainly from 2008 to 2018. The majority of respondents reported a rise in the revenues of commercial activities including cafés and restaurants, which they attributed to tourists' higher purchasing power and willingness to pay, causing on the other hand a struggle for the local population, which cannot afford the rise in housing prices and is forced to move to the neighbouring areas.

Despite the number of tourists in the historical centre and their economic effect on the rise of prices, tourism is mainly perceived as rather positive. In fact, most respondents recognized a general improvement in the city due to tourism development, including its level of attractiveness, even mentioning an increase in international recognition. They depicted the development as not only referring to better preservation of the historical centre but also recognized an improvement in the environment in terms of rebuilding programmes and cleanliness, and the enhancement of local culture. They felt proud of their city, and this was reflected in feelings of increased cultural identity and an overall attitude of hospitality.

### 7.5.2 The City of Milan

In Milan, host–guest encounters occur occasionally, but when they take place, they are mainly perceived as pleasant. In particular, residents appreciate the friendliness and the nice behaviour of tourists. Milan respondents named tourist behaviour as decisive for the quality of their relationships with foreign visitors, referring to disrespectful tourist attitudes as the cause of negatively perceived interaction. Although not reporting any direct negative experience, some respondents referred to vandalism as a negative tourist behaviour.

As far as changes in the urban setting are concerned, the interviewees also depicted a general improvement thanks to the regeneration programmes implemented by the municipality in recent years, especially on the occasion of the International Exposition in 2015. These programmes are perceived as a shift for Milan from a city towards a metropolis, comparing the city with London, Paris and Tokyo in terms of attractiveness, enhanced atmosphere and quality of life for both tourists and inhabitants, as they feel more attached to their city.

Different from Porto, Milan residents noted a scattered development of the city associated with tourism, in which the historical centre appears to be static but always well kept, while other areas are arising and establishing as new meeting centres, such as Isola and CityLife. Some respondents reported brand investments as relevant factors in urban change: they appreciated most the openings of a couple of international companies' stores in the historical centre, which have included processes of preservation and regeneration of the surrounding environment. Residents agreed on the economic benefits that foreign investors have brought to Milan, who are recognized as important tourism stakeholders. As a consequence, there is a perceived increase in business tourism demand and a rise in jobs, even though they are seasonal. In this context, some respondents also portrayed the struggle of local entrepreneurs with non-local companies that enter the market as a challenge. Besides the economic effects generated by foreign investors, Milan residents described the consequences of tourism in urban settings mostly positively and reported overall economic growth through the increase in tourism demand, as well as an increase in services and public transport, giving the example of the current construction of a metropolitan line with its destination Linate Airport. However, some described the city as chaotic, especially due to the traffic jams, perceiving in this regard also a worsening in the quality of the air. While on one hand, the interviews revealed a consensus regarding the increase of cultural offers and events, on the other hand, most residents perceived a loss in local identity, traditions and values and, since Milan is transforming into a metropolis, and so becoming rather international, they are dealing with multiculturalism.

## 7.6 Discussion and Conclusions

Host–guest interaction represents a fundamental dimension of the tourist experience in cities, since both tourists and residents are users of the same space and share the same facilities. The present study confirms the importance of the role of this social dimension in cities from the locals' perspective and that social interactions influence the residents' attitudes, opinions and lifestyles (Sharpley, 2008) in the cities of Porto and Milan. It is important to note that tourism in Porto is much more recent than in Milan and has helped the city overcome much of the financial crisis that struck Portugal hard in 2008. Additionally, the historical centre's regeneration is quite recent, so that people still recall quite vividly how much Porto has changed, while in Milan such changes have occurred over a longer period of time, and tourists and foreign (also business) visitors have been a common feature of daily life in the city for some decades. Here, also, a more scattered, tourism-shaped, urban development is visible, somehow relativizing the impact on the historical centre.

Perhaps due to this distinct context of tourism and place development (de Kadt, 1979), host–guest encounters occur more frequently and are more recalled as a significant part of daily life in the historical centre of Porto than in Milan. Social interaction was also confirmed, as correlated with the residents' job dependence on tourism; therefore, in accordance with Krippendorf's (1987) and Sharpley's (2014) research results, respondents in Porto interacted more frequently with tourists than those in Milan, since they have a higher proportion of job dependence on tourism compared to the participants in Milan. When those interactions occur, residents mostly perceive them as positive and friendly. The majority of interviewees, indeed, had never experienced really negative interactions, with few reporting disrespectful attitudes of tourists. Social relationships appear to be subjectively interpreted and linked to concrete tourist behaviours, and also to personal sensitivity to crowding. The majority of contacts with tourists are rather short, superficial street-interactions or take place in public spaces, at the metro station or in restaurants and stores, mainly initiated by tourists asking for help, coming close to Sharpley (2014) category of 'unintentional spontaneous encounters'. Since tourists are temporarily living in the cities, such encounters are, indeed, essentially transitory, non-repetitive and asymmetrical, in the sense that the meanings of these encounters differ for each party. For the local populations interviewed in Porto and Milan, such interactions with tourists are highly valued when they turn into pleasant conversations, getting to know each other, or living moments of entertainment. It is interesting to note that when interactions occur more frequently they are likely to have a high impact on residents' perceptions, positively affecting their degree of satisfaction. The results suggest that the overall experience of interaction with tourists is mostly positive and also emotionally rewarding in both cities. Similar results were found by Eusébio and Carneiro (2012b). The experience has been particularly positive while sharing meals, engaging in recreational activities, taking photos, inviting tourists to stay at home and participating in cultural activities, although closer interactions at home tend to occur less frequently.

The present study confirms that diverse residents hold distinct impact perceptions and attitudes towards tourism, as identified by several authors (Andriotis and Vaughan, 2003; Andereck *et al.*, 2005; Nunkoo *et al.*, 2013; Almeida García *et al.*, 2015), sometimes associated with distinct job dependency on tourism, i.e. corresponding to SET (Nunkoo *et al.*, 2013). Interviews in Porto reveal a consensus regarding the transformation towards tourism development, even though respondents perceived it differently, with some highlighting positive economic and social consequences (e.g. lively streets), while others focused on the negative impacts (prohibitive housing prices and crowding).

In the case of Milan, the residents' overall positive attitudes are related to a general transformation of the city, which is perceived as an improvement of the urban setting. Despite the differences of interpreting the urban changes through tourism development, both cities seem to have experienced a similar level of growth of international recognition, as expressed by their residents. This is probably related to the general improvement of both cities in terms of attractiveness thanks to regeneration programmes implemented by both municipalities. Interestingly, in Milan, the city has experienced a scattered development of its urban context, after an important international event in 2015, through which some peripheral areas are arising as new meeting centres, while the historical centre is still efficiently maintained, despite some degree of gentrification.

As far as residents' perceptions of the impacts of tourism are concerned, the residents' attitudes have been analysed by several authors according to the TBL approach (Eusébio *et al.*, 2014). Accordingly, residents hold positive attitudes concerning the economic benefits of tourism, especially regarding the increase in employment and job opportunities; however, they report negative consequences in the rise of prices and local taxes. In relation to the sociocultural dimension, besides residents' negative attitudes concerning a rise in crowding, the increases in cultural activities, public services and safety are generally perceived as positive consequences of tourism. As for environmental effects, local residents strongly valued a better preservation of historical buildings, urban spaces and green areas as a positive consequence of tourism, while reporting the environmental effects of noise and sound pollution, as well as decreased quality of air and water, also partly associated with increased tourist flows in both cities, even with slightly distinct contextual arguments.

Although the case studies presented confirm previous findings about the importance of social interactions for positively lived tourism experiences in cities and their role in conditioning residents' perceptions of the impacts of tourism, no statistically significant association between interaction and impact perception were identified. The limited dimensions of the sample studied here naturally limit the study itself. In addition, the case study methodology provided a focus on the historical centres of Porto and Milan, impeding the generalization of the results to other contexts. In order to overcome these limitations, similar empirical studies with a larger number of responses, eventually also including supply agents and visitors to both cities, in an attempt to understand distinct views from diverse but interacting stakeholders (Eusébio *et al.*, 2014), would be helpful in order to compare the insights from each urban environment and permit enhanced data exploration to identify more differentiated patterns among stakeholders.

In light of the Covid-19 pandemic, which had led to a drastic decrease in tourist flows, particularly to urban short-break destinations like Porto and Milan, populations may today respond differently to the questions posed here. Some questions should probably be added to reflect increasing risk-awareness in terms of propagation of disease through global travel, while the enforced lack of the mentioned positive, mainly economic, impacts might today lead to distinct overall evaluations, eventually outweighing perceived risks. How challenges and opportunities will be perceived in the post-Covid-19 period may explain residents' redefined view of tourism and tourists as well as future investments and projects. All these evolutions call for continuous research asking new questions and preparing new solutions. In times marked by increasing uncertainty, these solutions will also be context-shaped, as communities' views and attitudes rely on specific past experience of tourism and its impacts, which may, in the present case, lead to additional differences between Milan and Porto due to the distinct severity of the pandemic lived in the two cities. A follow-up study would therefore be particularly interesting.

## References

Adamiak, C. (2018) Mapping Airbnb supply in European cities. *Annals of Tourism Research* 71(12), 67–71. DOI: 10.1016/j.annals.2018.02.008.

Almeida García, F., Balbuena Vázquez, A. and Cortés Macías, R. (2015) Resident's attitudes towards the impacts of tourism. *Tourism Management Perspectives* 13, 33–40. DOI: 10.1016/j.tmp.2014.11.002.

Andereck, K.L., Valentine, K.M., Knopf, R.C. and Vogt, C.A. (2005) Residents' perceptions of community tourism impacts. *Annals of Tourism Research* 32(4), 1056–1076. DOI: 10.1016/j.annals.2005.03.001.

Andriotis, K. and Vaughan, R.D. (2003) Urban residents' attitudes toward tourism development: the case of Crete. *Journal of Travel Research* 42(2), 172–185. DOI: 10.1177/0047287503257488.

Ashworth, G. and Page, S.J. (2011) Urban tourism research: Recent progress and current paradoxes. *Tourism Management* 32(1), 1–15. DOI: 10.1016/j.tourman.2010.02.002.

Bimonte, S. and Punzo, L.F. (2016) Tourist development and host–guest interaction: An economic exchange theory. *Annals of Tourism Research* 58(4), 128–139. DOI: 10.1016/j.annals.2016.03.004.

Butler, R. (1980) The concept of a tourism area cycle of evolution. *Canadian Geographer* 24(1), 5–12.

Câmara Municipal do Porto (2011) *Censos 2011 – Mudanças demográficas*. March 2014, Municipality of OPorto, Porto, Portugal.

Chang, T.C., Huang, S. and Savage, V.R. (2004) On the waterfront: Globalization and urbanization in Singapore. *Urban Geography* 25(5), 413–436. DOI: 10.2747/0272-3638.25.5.413.

de Kadt, E. (1979) *Tourism: Passport to development?* Oxford University Press, New York, USA.

Doxey, G.V. (1975) A causation theory of visitor–resident irritants: Methodology and research inferences. In: *The impact of tourism: Travel Research Association*. 6th Annual Conference Proceedings, San Diego, September 8–11, pp. 195–198.

European Best Destinations (2017) Top destinations in Europe. Best places to travel in 2017. Available at: https://www.europeanbestdestinations.com/best-of-europe/european-best-destinations-2017/ (accessed 10 May 2019).

Eusébio, C.A. and Carneiro, M.J.A. (2012a) Determinants of tourist–host interactions: An analysis of the university student market. *Journal of Quality Assurance in Hospitality & Tourism* 13(2), 123–151. DOI: 10.1080/1528008X.2012.645201.

Eusébio, C. and Carneiro, M.J. (2012b) Socio-cultural impacts of tourism in urban destinations. *Revista Portuguesa de Estudos Regionais [Portuguese Review of Regional Studies]* 30(1), 65–76.

Eusébio, C., Kastenholz, E. and Breda, Z. (2014) Tourism and sustainable development of rural destinations: a stakeholders' view. *Revista Portuguesa de Estudos Regionais* 36(2), 13–21.

Freytag, T. (2009) Low cost airlines – A booster for European city tourism? [Low-Cost Airlines – Motoren für den Städtetourismus in Europa?]. *Geographische Rundschau* 61(2), 20–26.

Galdini, R. (2019) Urban re-use practices in contemporary cities: experiences in Europe. *Cities* 87(2), 103–105. DOI: 10.1016/j.cities.2018.12.026.

García, B. (2004) Cultural policy and urban regeneration in Western European cities: Lessons from experience, prospects for the future. *Local Economy* 19(4), 312–326. DOI: 10.1080/0269094042000286828.

GaWC - Globalization and World Cities Research Network (2010) The World According to GaWC 2010, Classification of cities 2010. Available at: https://www.lboro.ac.uk/gawc/world2010t.html (accessed 10 May 2019).

GaWC - Globalization and World Cities Research Network (2018) The World According to GaWC 2018, Classification of cities 2018. Available at: https://www.lboro.ac.uk/gawc/world2018t.html (accessed 10 May 2019).

Hayllar, B., Griffin, T. and Edwards, D. (2008a) Urban tourism precincts: Engaging with the field. In: Hayllar, B., Griffin, T., and Edwards, D. (eds) *City Spaces, Tourist Places: Urban Tourism Precincts*. Routledge, Amsterdam, The Netherlands, pp. 3–18.

Hayllar, B., Griffin, T. and Edwards, D. (eds) (2008b) *City spaces, tourist places: Urban tourism precincts*. Routledge, Butterworth-Heinemann, Amsterdam, The Netherlands.

INE (2017) *Estatísticas do Turismo 2017 [Tourism Statistics 2017]*. INE, Lisbon/ Portugal.

Ioannides, D., Röslmaier, M. and van der Zee, E. (2019) Airbnb as an instigator of 'tourism bubble' expansion in Utrecht's Lombok neighbourhood. *Tourism Geographies* 21(5), 822–840. DOI: 10.1080/14616688.2018.1454505.

ISTAT (2018) Dati ISTAT, Popolazione residente al 1° Gennaio 2018, Lombardia. Milano. Available at: http://dati.istat.it/Index.aspx?DataSetCode=DCIS_POPRES1# (accessed 10 May 2019).

Kastenholz, E. (2010) 'Cultural proximity' as a determinant of destination image. *Journal of Vacation Marketing* 16(4), 313–322. DOI: 10.1177/1356766710380883.

Kastenholz, E., Carneiro, M.J., Eusébio, C. and Figueiredo, E. (2013) Host–guest relationships in rural tourism: Evidence from two Portuguese villages. *Anatolia* 24(3), 367–380. DOI: 10.1080/13032917.2013.769016.

Krippendorf, J. (1987) *The Holiday Makers: Understanding the Impact of Leisure and Travel*. Elsevier Science & Technology Books, University of Minnesota/ USA.

Law, C. (1996) *Tourism in Major Cities*. International Thompson Business Press/Routledge, London/ UK.

lidisSAS Visual Analytics Viewer (2015) Turisti stranieri per provenienza e struttura, Comune di Milano [Foreign tourists by origin and structure, Municipality of Milan]. Available at: http://www.comune.milano.it/wps/portal/ist/it/amministrazione/datistatistici/sisi2/commercio_turismo/turismo (accessed 10 May 2019).

Namberger, P., Jackisch, S., Schmude, J. and Karl, M. (2019) Overcrowding, Overtourism and Local Level Disturbance: How Much Can Munich Handle? *Tourism Planning & Development* 16(4), 452–472. DOI: 10.1080/21568316.2019.1595706.

Ngowi, R.E. and Jani, D. (2018) Residents' perception of tourism and their satisfaction: Evidence from Mount Kilimanjaro, Tanzania. *Development Southern Africa* 35(6), 731–742. DOI: 10.1080/0376835X.2018.1442712.

Nunkoo, R., Smith, S.L.J. and Ramkissoon, H. (2013) Resident's attitudes to tourism: a longitudinal study of 140 articles from 1984 to 2010. *Journal of Sustainable Tourism* 21(1), 5–25. DOI: 10.1080/09669582.2012.673621.

Peris-Ortiz, M., Bennet, D.R. and Pérez-Bustamante Yábar, D. (2017) *Sustainable Smart Cities: Creating Spaces for Technological*. Social and Business Development, Springer International Publishing, Switzerland.

Ray, C. (1998) Culture, intellectual property and territorial rural development. *Sociologia Ruralis* 38(1), 3–20. DOI: 10.1111/1467-9523.00060.

Richards, G. (2000) The European cultural capital event: Strategic weapon in the cultural arms race? *International Journal of Cultural Policy* 6(2), 159–181. DOI: 10.1080/10286630009358119.

Richards, G. (2014) Creativity and tourism in the city. *Current Issues in Tourism* 17(2), 119–144. DOI: 10.1080/13683500.2013.783794.

Sharpley, R. (2006) *Travel and Tourism*. SAGE, London.

Sharpley, R. (2008) *Tourism, Tourists and Society*, 4th edn. Elm Publications, Huntingdon, UK.

Sharpley, R. (2014) Host perceptions of tourism: A review of the research. *Tourism Management* 42(7), 37–49. DOI: 10.1016/j.tourman.2013.10.007.

Stevenson, D. (2003) *Cities and Urban Cultures*. Philadelphia, Open University Press.

Stylidis, D., Biran, A., Sit, J. and Szivas, E.M. (2014) Residents' support for tourism development: The role of residents' place image and perceived tourism impacts. *Tourism Management* 45(7), 260–274. DOI: 10.1016/j.tourman.2014.05.006.

United Nations, Department of Economic and Social Affairs, Population Division (2019) *World Urbanization Prospects 2018: Highlights (ST/ESA/SER.A/421)*. UNWTO, Madrid/ Spain.

UNWTO (2002) *Tourism 2020 Vision – Global Forecast and Profiles or Market Segments*, Vol. 7. UNWTO, Madrid/ Spain.

UNWTO - United Nations World Tourism Organization (2012) *Global Report on City Tourism – Cities 2012 Project*. UNWTO, Madrid/ Spain.

Wine Tourism Portugal (2017) Porto Best European Destination 2017. Available at: https://blog.winetourismportugal.com/porto-best-european-destination-2017 (accessed 11 May 2019).

Wise, N. (2016) Outlining triple bottom line contexts in urban tourism regeneration. *Cities* 53, 30–34. DOI: 10.1016/j.cities.2016.01.003.

# 8 Sociocultural Dimensions of Destination Resilience and Implications for Innovative Product Design in Experimental Destinations in Germany, Italy and Ireland

**Valeria Diaferia[1]*, Anastasia Traskevich[2] and Martin Fontanari[3]**
*[1]Hochschule Fresenius University of Applied Sciences, Cologne, Germany; [2]Belarusian State Economic University, Minsk, Belarus; [3]International School of Management (ISM), Cologne, Germany*

## 8.1 Introduction

The ongoing global economic, environmental and social changes brought by the crisis of Covid-19 have had a huge impact on tourism and, more specifically, on tourism destinations in Western Europe. Therefore, considering the post-Covid-19 world as the central issue of theoretical and practical application, the challenge for the tourism destinations in the region is to re-evaluate the new macro- and microeconomic trends and to proceed towards a long-term recovery and innovation-based resets.

In order to cope with complex and interrelated change impacts (Luthe and Wyss, 2014) and make destinations concentrate more on their ability to adapt to changing conditions and circumstances, researchers introduced a new perspective: they thematized the quality of resilience (Davoudi and Porter, 2012), as applied to the various components and dimensions of tourism. The present study addresses the paradigm of destination resilience, a current trend orientation in tourism studies (Dahles and Susilowati, 2015; Hall *et al.*, 2018; Innerhofer *et al.*, 2018; Pechlaner *et al.*, 2019; Ritchie and Jiang, 2019; Romão, 2020; Fontanari *et al.*, 2021; Gössling *et al.*, 2021), with the practical aim of developing an adequate response to the challenges of tourism development in Western Europe and to reinvent approaches to product design. Due to its dynamic nature, a new topical field of investigation is the specific dimension of *sociocultural resilience*. The sociocultural dimension of destination resilience for sustainable and responsible development of tourism is explored through this chapter as an important aspect of destination resilience and authentic product design. Furthermore, sociocultural resilience as a theoretical concept is practically investigated within its implications for innovative product design in experimental tourism destinations in Germany, Italy and Ireland.

The basis of the sociocultural resilience approach is implemented within a 'Code-Red Model' (Traskevich and Fontanari, 2018). The system of qualitative and quantitative indicators is introduced in the current study as a methodological approach to investigate sociocultural

*Corresponding author: valeriadiaferia7@gmail.com

 *Tourism Planning and Development in Western Europe* (eds K. Andriotis *et al.*)
DOI: 10.1079/9781800620797.0008

resilience in the aforementioned countries. For the methodological framework, this model has been chosen because of its orientation towards personal resilience, because of the complexity of the dimensions the model describes, the resilience of the community, the accent on respectful and well-minded co-inhabitation of residents and guests as well as sustainable mutual influence of people and nature, and its authentic approach in terms of intensive regional development.

The indicators, based on the aforementioned Code-Red Model, cover the following characteristic dimensions of tourism communities: emotions and creativity; learning skills and professional efficiency; inner and outer world balance; communication and positive thinking; ethics and values; and spirituality. These indicators are applied to implement the comparative analysis on the basis of collected secondary data with further assessment of sociocultural resilience in its perspective application to tourism for the selected countries. The comparative analysis allows the suggestion, in the last part of this chapter, that innovative and resilience-oriented tourism products contribute to advanced sustainable and resilient tourism development in Western Europe, with a particular focus on enhancing experimental destinations in their positive response to change.

## 8.2 Theoretical Framework to Investigate the Sociocultural Dimension of Destination Resilience

Until the heavy impacts of Covid-19, global tourism growth was unprecedented. Consequently, this has elevated the sector as a key driver of global socioeconomic progress and has led the tourism sector as a representative social and economic system which defines the strategic orientation of a region (Cochrane, 2010). However, complex trends as well as political, economic, environmental and social developments, and permanent and temporary disturbances, endanger destinations and influence the market, the current business models of tourism enterprises and product orientation.

Tourist destinations can face a range of change pressures: they may be environmentally induced (e.g. impact of natural hazards and global environmental change) or socially induced (e.g. changing cultural resources), or they may experience severe challenges of an economic nature (e.g. changes in economic conditions, loss of tourist markets) (Lew, 2014). At the same time, the devastating recent effects of the Covid-19 pandemic illustrate the vulnerability of the global economy to the events of systemic disorders. In critical situations, like the pandemic outbreak, tourism destinations appeared totally locked down and unable to come back to their initial authentic way of life and economic activities.

Moreover, ongoing globalization, the emerging digitalization of the state, economy and society, as well as the demand for global measures to avert climate change, can be considered as further impacting events on the destination side. Furthermore, critical phenomena such as terrorism can be included as part of the changes and crises in tourism. In fact, political unrest and terrorist attacks are problematic for a tourism industry that reflects a myriad of experiences that it sells to consumers (Richter and Waugh, 1986; Mansfeld, 1999). These contemporary global change processes, increasing uncertainty and insecurity, as well as economic and environmental shocks and the growing awareness of the negative implications of naturally- or human-induced developments, have led to a higher sense of sensitivity to the ability or capacity of a system to deal with sudden and unexpected changes (Sheppard and Williams, 2016). In fact, in the last two decades, more and more articles have been published to address the context of crises and conflicts affecting tourism. The vulnerability of destinations to shocks, stressors and sudden changes has become a rising concern for researchers and industry stakeholders (Luthe and Wyss, 2014).

In order to deal with the question of crisis in modern framework conditions, researchers have introduced a new perspective within the concept of resilience (Davoudi and Porter, 2012) as applied to tourism. It is most often discussed in tourism research as the capability of different components of a tourism system to overcome and positively react to crises or disturbances of a different nature. In fact, tourism appears to be an appropriate field for applied research on resilience, just as the tourism destination appears to be an appropriate level at which to observe resilience (Partale and Traskevich, 2018). From this perspective arises the concept of destination resilience which can be understood as the ability of a destination to be prepared in the best possible way for disruptions and crises and to provide the minimum basic services within a tourism context or to guarantee them with as little delay as possible (Fontanari and Ewald, 2018). Hence, tourism destinations can be seen as an open regional system influenced and surrounded by diverse aspects and players who contribute to a destination's functional autonomy, safety, equilibrium, uniqueness, sustainability, flexibility and adaptability to crises (Traskevich and Fontanari, 2021).

According to the academic literature, these dimensions can be summarized in three categories to describe and discuss the resilience of a destination: the ecological–environmental dimension, the economic–fiscal dimension and the social–cultural (sociocultural) dimension (Calgaro *et al.*, 2013).

The sociocultural dimension embraces three concepts: society, culture and system. From a tourism destination perspective, a 'society' can be seen as formed from the relations between locals and guests. Culture is the learned behaviours that are shared by the members of a society/community, together with the material products of such behaviours. Learned behaviour includes knowledge, beliefs, values, customs and art acquired by a person as a member of a society. Material products or objects are represented by the tourist products which are part of a tourism destination, aimed at tourists and constituting a representation of a people's way of life (Kauffman, 1980). The words 'society' and 'culture' are fused together to form the word 'sociocultural'. The term is used to give a holistic approach to looking at non-material cultural constructs (such as faith, value systems and epistemology) and social structures (such as political organizations, civil organizations, educational systems, etc.), and at how the two interact.

Therefore, the sociocultural dimension can include an individual's subjective experience, which is represented by the emotional experience and psychological frameworks of the individual, which are then shared by a community/tourism destination through an exchange or material expression of the culture, language, artefacts and rituals. Hence, it is possible to summarize the key conceptual characteristics of the sociocultural dimension on the basis of the destination-clustering approach (Traskevich and Fontanari, 2019) in the following way: cultural values, educational systems, social behaviour, faith and spirituality, emotions and creativity (Traskevich and Fontanari, 2018). The challenge related to sociocultural resilience is then to understand the essence and dimensions of a healthy society and the role and the place of personal health within this society. In this respect, the local population is not only a background artist having a positive attitude toward tourism but has to be involved actively in a creative way to co-create future-oriented tourism development themes (Beeton, 2006). Therefore, sociocultural resilience to stress on local people and communities at a destination is particularly related to the mental ability and skills of locals and guests to deal with crises.

## 8.3 Methodological Approach to Assess Sociocultural resilience for Tourism Destinations

Thanks to the Convergent Development Model of Mental Wellness within resilient destinations (the Code-Red Model, presented and developed by Traskevich and Fontanari, 2018), new mental wellness products can be developed in order to contribute to sociocultural and destination resilience. The sources that are going to be used for the methodology are parts of the Code-Red Model, which has the particular aim of analysing (i) the development path to resilient personality, describing personal mental wellness from an individual perspective and (ii) the development path to resilient destinations, describing destination resilience from a regional perspective (Traskevich and Fontanari, 2018, 2019). Personal mental wellness is well known for its emphatic awareness of mental and spiritual health. Hence, the sources used to describe the individual perspective are based on the Code-Red Model and the following: emotions and creativity, learning skills and professional efficiency, inner and outer world balance, communication and positive thinking, ethics and values, and spirituality.

Furthermore, personal mental wellness is often described as a state of consciousness which cannot be viewed separately from the environment in which it occurs (Cohen and Bodeker, 2009) – that is, destination/community in the context of tourism. For this reason, it is possible to identify the corresponding dimensions/sources for the concept of destination resilience. According to Traskevich and Fontanari (2021), these functional components are the following: the unique nature of the tourism product, its functional autonomy, a resilient tourism product value chain, safety, consistency and equilibrium, sociocultural resilience and resilient sustainability.

The current study focuses on comparative assessment of sociocultural resilience for three different countries of Western Europe: Germany, Italy and Ireland. The countries are chosen because of the practice-oriented idea which bases itself on the perceived immense potential to develop experimental resilient destinations there. At the same time, assessing sociocultural resilience and understanding the potentials of the sociocultural dimension of destination resilience promises, in the case of the present country selection of Germany, Italy and Ireland, the unique and authentic diversity of expected results due to obvious and quite iconic sociocultural differences in these three countries. Thus, the perspectives for comparison, typologizing and clustering become of particular academic interest due to the chosen selection of the countries for investigation and analysis. The previously described model is applied within the current methodology as a framework of indicators through which sociocultural resilience can be described, compared and assessed on the basis of secondary data for the three countries.

- **The first indicator of sociocultural resilience** – converging emotion and creativity and unique selling propositions (USPs) – concerns itself with the desire that both an individual and a tourism destination show their best qualities. From this consideration, the two suggested indexes, proposed respectively from the individual and destination perspectives, and then used to investigate and assess sociocultural resilience for the three countries, are: (i) *evaluating national pride and analysing the natural and cultural heritage*; and (ii) *uniqueness through recognized World Heritage Sites*.
- **The second indicator** – learning skills, professional efficiency and functional autonomy – is related to the tendency that humans have to constantly improve their learning skills, competencies and experiences, becoming efficient in practising their daily tasks. The cognitive capacities or 'toolkit' intended as a range of competencies, flexibility/resilience, and funds of knowledge (Swidler, 1986) enhances intellectual skills which are useful in problem-solving in a wide array of topics and issues, improving the academic performance of individuals. People who achieve a given level of education, regardless of how they got there, have toolkits that are in a similar range of value to employers. This approach suggests that a substantive connection between learning skills and performance in school or university accounts for a high effect on educational attainment. Then, at the next step of the process, education has a substantial payoff in occupational status. Accordingly, two suggested indexes to be analysed for the three countries are: (i) *analysing the variety and quota of tertiary education graduates*; and (ii) *employment rates*. Destinations perform resiliently if they apply to regional resources to design attractive and unique tourism products. *Rating the handicraft sector*, in particular, the number of handicrafts employed in the selected countries can, thus, be considered in this context as a third suggested index. In recent years, an increase in the concern about local ways of life (e.g. festivities), has motivated the conservation of intangible cultural heritage specifically, and contributed to the continuity of the implementation of traditional craftsmanship as a domain of intangible cultural heritage which can also been seen as a valuable resource for the development of resilient and unique tourism products.
- **The third indicator** – inner and outer world balance and resilient value chain – refers to the ability that humans have to react positively when challenged with new situations or critical circumstances. In this sense, an index suggested to measure

this indicator in the selected countries is *rating the happy nation index*, as a way to rank the national happiness of people based on respondent ratings of their own lives, correlated with various life factors. Consequently, when it comes to the destination, a juxtaposition of many individuals' self-sufficient structures must be intensively linked in a synergic way. For instance, some experiences such as participation in events involve a presence of both local and guest. In fact, the *attendee event index* is proposed to investigate the three countries from a destination perspective.

- **The fourth indicator** – communication, positive thinking, safety, consistency and equilibrium – refers to the ability of people to communicate with other people. Families play a central traditional role in societies, as they are the primary site for reproduction and in socializing the next generation. Following this consideration, two indexes can be suggested as representative of communication, safety and stability from an individual perspective: (i) *the proportion of households with children*; and (ii) *crude divorce rate*. It comes to light that stable people form stable structures, which within a destination leads to a more stable local community. Tourists and guests tend to communicate and share their feelings or opinions with others who are completely involved in a place, for instance clubs/associations. Hence, rating the *number of cultural and sports clubs/associations in relation to the population* as opportunities for social interaction through which it is possible to develop the knowledge, skills and attitudes necessary for a full participation in civil society, is suggested as an index from the destination perspective, which will be further analysed for the three countries.
- **The fifth indicator** – ethics, values and sociocultural resilience – highlights the importance of ethics and values as preserved by culture and tradition over centuries. Following this, the *number of local/regional museums* is presented in this context to conduct the comparative analysis for the three countries. Museums play a crucial role as guardians of values and heritage of a cultural (historical) location (destination) and they have a duty to preserve and protect artefacts for the future, enabling current and future generations to become familiar with their heritage.
- **The sixth indicator** – spirituality and resilient sustainability – focuses on the preserving and caring impulse of a person who prefers spiritual goals to material ones. On the destination side, sustainability receives a new meaning in terms of the individual's ability to adapt patiently towards vulnerable situations and restabilize human living conditions. In this context comes to light the concept of religious tourism as a type of tourism exclusively or strongly motivated by religious reasons. A practical research focus for the three countries can therefore be measuring the *participation in activities of churches* as a faith representative and the number or quantity of some spiritual tourism activities such as special *pilgrimage routes*. To analyse the presented six indicators and the aforementioned measuring indexes for each of the indicators, substantial statistics, national reports and the quantitative results of research projects are investigated in search of the relevant statistical data. All the statistical parameters are normalized using the usual index approach to ensure data comparability.

## 8.4 Empirical Results: Comparative Analysis and Assessment of Tourism Potentials of Sociocultural Resilience in Germany, Italy and Ireland

The empirical results of comparative analysis and assessment of parameters of sociocultural resilience in Germany, Italy, and Ireland are summed up in Table 8.1.

Referring to the first suggested *national pride index*, part of the first indicator, the most comprehensive data set on national pride is the ISSP-module (International Social Survey Program) designed by Haller and Yuchtman-Yaar (2009) in which national pride is investigated in terms of the country's science and technology, economy, arts and literature, and sport. Measuring the degree of national pride for the three selected

**Table 8.1.** Comparative analysis and assessment of indicators of sociocultural resilience in Germany, Italy and Ireland.

| Indicator of analysis | Index for assessment | Germany | Italy | Ireland |
|---|---|---|---|---|
| Emotions and creativity of a person vs USP of a destination | Science and technology | 64 points | 59 points | 63 points |
| | Pride in arts and literature | 66 points | 80 points | 83 points |
| | Pride in sports | 64 points | 67 points | 85 points |
| | Pride in economy | 66 points | 37 points | 67 points |
| | UNESCO world heritage: cultural sites | 43 sites | 49 sites | 2 sites |
| | UNESCO world heritage: natural sites | 3 sites | 6 sites | – |
| Learning skills, professional efficiency and functional autonomy | Variety and quota of tertiary education graduates | Engineering, manufacturing, construction fields: 22%; graduates in health and welfare: below 8%; natural sciences, mathematics, statistics and ICT: 14% | Arts and humanities: 16%; education field: below 5% | Engineering, manufacturing, construction fields: below 10%; natural sciences, mathematics, statistics and ICT: 15% |
| | Employment rates | 91% | 52% | 84% |
| | Rating the handicraft sector | Craft market in 2019 had decreased by more than 11% over the previous 5 years from €1.306 million in 2014 to €1.159 million in 2018 | Small enterprises: 13%; 78% of companies are more vulnerable than other businesses | 11,415 people employed; the sector makes an important contribution to economic development; potential for growth |
| Inner and outer world balance and resilient value chain | Happy nation index | 17th position with 6.985 score on 156 countries | 36th position with 6.223 score | 16th position with 7.021 score |
| | Attendee event index | 77.2% of the population aged 16 or over in 2015 attended at least one cultural or sport event | 49.6% | 77.2% |

*Continued*

**Table 8.1.** Continued

| Indicator of analysis | Index for assessment | Germany | Italy | Ireland |
|---|---|---|---|---|
| Communication and positive thinking vs safety, consistency and equilibrium | Number of cultural and sports clubs/ associations in relation to the population | 246 cultural associations; 89,000 sports associations (27 million members); 939 people per 1 association | 192 cultural associations; 95,000 sports associations; 635 people per 1 association | No data available about the number of cultural associations; 12,000 sports associations; 411 people per 1 association |
| | Proportion of households with children | 22% proportion of households with children | 8% proportion of households with less than 3 children | 41% proportion of households with children |
| | Crude divorce rate | 2.3% | 1.5% | 0.7% |
| Ethics, values and sociocultural resilience | Number of local/ regional museums | 6710 museums (2,586 local/regional managed museums) | 4976 museums (1480 local/regional managed museums) | 230 museums (23 local/regional managed museums) |
| | Participation in activities of churches | 16.8% | 20.8% | 50.3% |
| | Number of pilgrim routes | 1600 | 397 | 9 |

countries shows the following results. Ireland and Germany are prouder in regard to science and technology than Italy. The Irish are proudest in regard to arts and literature, sports and in the economy of their country. Also, Italians are very proud in regard to arts and literature and sports; however, they aren't very proud in terms of their economy. Ireland obtains the highest score of national pride in economy immediately followed by Germany.

The second index of the first indicator focuses on *analysing uniqueness through recognized World Heritage Sites*. An analysis was made for the three selected countries according to the UNESCO World Heritage Centre. Italy has the largest number of World Heritage Sites (55), and in terms of these being representative of uniqueness, Italy has the comparative advantage for tourism development in terms of its cultural and natural attractions and cultural heritage over Germany and Ireland. However, Germany comes next due to its high number of World Heritage Sites. Even though Ireland is recognized from the previous research as a country with the highest national pride, when discussing its World Heritage Sites, it comes in the last position.

Applied considerations towards sociocultural resilience in the selected countries which can be derived from the study using the index *variety and quota of tertiary education graduates and employment rates* are the following. Germany and Ireland have the same number of students graduated in natural science, mathematics and statistics. In Germany, there is the highest quota of graduates in engineering, manufacturing and construction fields; whereas Italy prevails in the number of graduates in arts and humanities, which is confirmed by its flourishing cultural heritage. The highest employment rate is in Germany: a country with the highest number of graduates.

*Rating the number of handicrafts employed* in the three selected countries is complicated to carry out. It is hard to access reliable statistics about the sector (Craftsman Project, 2010; Indecon International Economic Consultant,

2010; Craft Industry Alliance, 2019; Eurostat, 2019, 2020b) as there aren't any national institutions which systematically collect data on the sector. Nevertheless, an attempt is made here to show a European picture, specifically related to the three selected countries. It can be stated that in Germany from 2014 to 2019 handicraft production decreased. In addition, in Italy, starting from 2009, the handicraft sector has been in a difficult economic situation due to economic, social and ecological changes. The study indicates a higher number of persons engaged in the craft sector in Ireland.

One of the two suggested indexes related to the indicator on inner and outer world balance and resilient value chain is the *happy nation index*. The study refers to the *World Happiness Report* (Helliwell *et al.*, 2019). Within this index, Ireland and Germany show typical life satisfaction. However, Italy, being at the 36th position, might reflect the largest decline in average life evaluations. This can be explained by a combination of economic, political and social stresses which became typical for the lifestyle in the country. People recognize that communal creativity and celebrations might be opportunities to react positively when critical situations occur. This locally-based creativity presents occasions for development in the dimensions of both sociocultural resilience and attractive destinations. When considering the second index, *attendee event index* (Eurostat, 2017b), Italy shows itself as a country with little inner and outer world balance and resilient value chain. This means that Italians, when compared to Germans and the Irish, have not yet got the right capacity to react positively when facing disruptive changes or difficult situations. This implies that Italy has difficulties in improving opportunities for the development of resilient structures at the level of a destination.

Also, the number of cultural and sports clubs/associations in Germany, Italy and Ireland is measured (Federation of Irish sport, 2017; German Olympic Sports Confederation, 2018) in relation to the population and is suggested as one of the three indexes referring to the fourth indicator, 'Communication and positive thinking and safety, consistency and equilibrium'. The proportion of households with children (Eurostat, 2017a) is expected to show the role that families play in societies as primary sites in the socializing and education of next generations. From this perspective and the data presented in Table 8.1, it follows that Ireland potentially represents a more stable country with more resilient people.

The third index analysed is the *crude divorce rate*. Here, Germany appears to be the country with the highest number of crude divorces compared to Italy, followed by Ireland (Eurostat, 2020a). This result is quite conflicting if compared to the previous outcomes coming from the other indexes.

Considering the indicator on ethics, values and sociocultural resilience and its importance as a source preserved by culture and traditions over centuries, the *number of local/regional museums* is suggested as an index as these institutions act as guardians of values and heritage for the three countries. The collection and administration of European data on museums are coordinated by the European Group on Museum Statistics (2020). Comparing the results for the three countries, it can be stated that Germany has preserved ethics, values and cultural heritage contributing to the overall resilience of the country. Germany also seems to promote an alternative lifestyle which can be integrated into the everyday life of a guest who gets in touch with the authenticity of the country (region) when visiting a museum. Italy comes next, due to the high importance that culture, heritage and values have for the country as revealed previously from some of the aforementioned indexes. Ireland comes in last position and this can be explained by the fact that the country is not very well developed in terms of designing profitable and attractive cultural touristic products.

The last two indexes refer to the last indicator of sociocultural resilience, 'spirituality and resilient sustainability', which focuses on people's ability to preserve spiritual goals. It follows the *participation in activities of churches* as an index to investigate for the three countries (Eurostat, 2006). For church and religious activities, a higher level of participation can be observed in Ireland. It follows that Ireland as a destination and the Irish as a people appear to adapt toward vulnerable situations more traditionally and spiritually than people in Italy and Germany thanks to faith, hope and optimism. The second index of this indicator investigates the number of spiritual activities such as the number of *pilgrimage walks* in the three countries. The results observed are in contrast with the previous findings. In fact, Germany is the country with the highest number of pilgrim routes which are deeply embedded in the protected natural landscapes of the country.

The brief analysis of the sociocultural characteristics of destination resilience conducted for the three selected countries under the framework of the Code-Red Model allows us to derive the key development trends as well as sociocultural gaps and potentials. These outcomes allow for practical consideration of designs for attractive, competitive and resilient tourism products for the experimental destinations in the selected countries.

## 8.5 Discussion: Product Development Solutions to Enhance Sociocultural Resilience in the Selected Countries

Thinking about resilience in tourism means understanding that it is not only a 'technical' (Holling, 1996) but also a geographical and spatial topic (Luthe and Wyss, 2014). In this way it becomes clear that increasing and ensuring resilience of tourism destinations is a more complex task than doing so for 'normal' systems and even cities (in comparison to countries as destinations) (Duit *et al.*, 2010). Considering solutions at the planning level, strategic considerations and product development concepts already lead one to consider simple ways of solving or preventing problems in advance (Uğur, 2018). In terms of product development solutions, product design should still be based on universal laws and approaches which are well established in tourism theory and practice (Walder *et al.*, 2005; Partale and Traskevich, 2018; Partale and Partale, 2019, 2020; Prieto, 2020; Tomej and Xiang, 2020). However, new insights should be creatively developed to gain the multidimensional quality of resilience (for tourists, local communities, landscapes, regions, economic clusters, etc.) through tourism activities and tourism economic development in destinations (Traskevich and Danskikh, 2018; Traskevich and Fontanari, 2018; Pechlaner *et al.*, 2019; Stienmetz *et al.*, 2020; Volgger *et al.*, 2021; Volo, 2021).

On the basis of the conclusions coming from the comparative analysis of the three countries and their indicators on sociocultural resilience, tourism is revealed as an important tool to enhance the sociocultural resilience of destinations. In particular, attention is focused on establishing and strengthening the tourism development strategies in these countries by concentrating on the core images, attractions, specific resources, abilities and product development solutions for the respective destinations. Product development suggestions are based on the outcomes of the comparative analysis and assessment of sociocultural resilience in Germany, Italy and Ireland and are systematized in Table 8.2.

### 8.5.1 Germany

According to the results of the analysis regarding the indicators of sociocultural resilience, Germany appears to give its visitors the image of a positive, stable and friendly country but also a place for business and investment overall. At the same time, the increasing shortage of skilled labour has become an extremely important topic both for the economy and for the tourism industry in Germany. The wide range of cultural attractions need not be limited to museums, theatres or castles in well-known destinations. Germany has a lot to offer in more remote rural areas (emerging or experimental destinations) where authentic traditions and customs can be explored by tourists. Developing tourism in rural areas, which are often economically weak regions, could also increase job opportunities and the income of these regions. In addition to this, increasing resilient tourism measures as the 'new normal' of the 'great reset' in the tourism industry might be a strategic solution to enhance sociocultural resilience in remote destinations, which in Germany comprise the greater opportunities to develop towards the concept of destination resilience.

Furthermore, medical and wellness tourism deserves, in this context, new considerations regarding innovative product orientation

**Table 8.2.** Development suggestions for the design of authentic and resilient tourism products for experimental destinations in Germany, Italy and Ireland.

| Country | Estimation of sociocultural resilience | Tourism potentials of sociocultural resilience | Product development priority suggestions | | | |
|---|---|---|---|---|---|---|
| | | | *Tourism theme 1* Design suggestions | *Tourism theme 2* Design suggestions | *Tourism theme 3* Design suggestions | *Tourism theme 4* Design suggestions |
| Germany | Culture is shared by the community together with its knowledge, beliefs, values, customs: high national pride in economy; intellectual skills as tools to enhance productivity and gain independence; manage connectivity | Historic ability to overcome crises | *Wellness tourism:* spiritual healing practices; positive thinking; learning experiences | *Edutainment:* rising awareness of local attractions; skills, knowledge; hands-on experience | *Personal sustainability:* involvement with nature | *Rich culture:* focus on local authenticity |
| Italy | More vulnerable to shock or stress: low national pride in economy; low employment rate; last position as happy nation | Preserve culture and tradition, ethics and values, cultural artefacts | *Discovery of small local attractivity*: skills and knowledge; success in economy; local resources as a way to gain independence | *Co-inhabitation of locals and guests:* communal creativity and celebrations; interactive events; coaching on positive reaction and life-wisdom | *Culture and traditions:* sustainable living; highlight local products in experimental destinations | *Spirituality:* wellness tourism products; religious tourism; rural tourism |
| Ireland | High national pride; highest happy nation index; high proportion of households with children; protect local jobs; preserve spiritual goals over material ones | Admirable quality in reacting positively when challenges occur | *Cultural heritage:* exploring rich potentials of experimental destinations; focus on local resources | *Wellness tourism:* spiritual wellbeing; personal development; reconnection to nature; family wellness and 'we-time' | *Sociocultural interaction:* interactive cultural events; activities within associations; cultural exchange between guests and locals | *Religious tourism:* pilgrimage; religious family activities and celebrations |

towards spiritual wellness packages aimed at achieving a sense of a healthy and meaningful life – which can be especially relevant after the Covid-19 crisis. Within the current research, the resilient-oriented product proposed for Germany is 'Detoxified in the forest'. The idea of this new tourist product is born from the consideration that Germany offers large areas covered by forest with the best tourism infrastructure available for outdoor recreation. Nature-based tourism is a constantly growing trend which was rediscovered in a completely different context during the Covid-19 crisis. Therefore, this new experiential theme can be identified as a priority suggestion for product design in nature-based destinations in Germany. The goal is to provide a brief exploration of tourist practices of forest-based tourism, which are contextualized in Germany, particularly, in relation to wellbeing, spirituality and adventure tourism.

Another aspect of this tourist product is the possibility to experience a real 'digital detox' holiday. The predominance of 'natural' resources (forests, lakes, rivers, wildlife), as well as the experiential activities related to detaching from human–urban areas are very important and well reflected in 'catch-up' tourism activities post-Covid-19. Involving all these elements and activities in resilient product design will enhance the sociocultural resilience of a region in a sustainable way, thus strengthening local tourism and cultural and human potential.

### 8.5.2 Italy

Concerning the considerations on product design coming from the conducted comparative analysis of sociocultural resilience, Italy appears to be more vulnerable to shock or stress. The following strategic suggestions aim to identify a new method of employing tourism in Italy. First among these is to develop and promote new (experimental) destinations, as well as to design resilient-oriented tourism products, in order to generate both economic and social benefits and reach the ability to overcome the post-Covid-19 crisis in a resilient way. First of all, the position of Italy's tourism sector and the potentials for new tourism destinations are both extremely high in terms of the appeal of cultural and environmental heritage, which remains a unique and distinctive competitiveness factor in the tourism market. However, Italy has difficulties in positioning itself on emerging (niche) market segments and even as a touristic country, in general, with regard to the new demand flows, innovative product trends, new risks and uncertainties created by development of geopolitical instability and the current impacts of the pandemic. This also influences Italy's positioning in terms of global reputation, especially from the perspective of the 'great reset' for the tourism industry post-Covid-19.

Within the new framework conditions created by Covid-19, Italy should exploit further its competitive potential in order to develop and promote resilient destinations. For instance, culture and tradition should be preserved in the long term in order to increase sociocultural resilience and sustainability. Tourist flow is still too heavily concentrated on several sites of extraordinary value, thus increasing the risk of loss of territorial identity. Interventions such as incrementing the accessibility to places and territories through sustainable mobility systems need to be reinforced in order to assist Italy in acquiring a new position of leadership based on tourism resilience, sustainability, innovation and competitiveness. Furthermore, more attention needs to be given to local products, by restarting and developing different locally-based sectors such as the manufacturing industry and agriculture, preserving local culture and traditions as distinctive elements of the regional brands.

Sustainable measures should focus on the redevelopment and recovery of the environment. For instance, rural and religious tourism can be considered with the applied consideration of developing religious routes and itineraries which involve close contact between nature and tourists, but also between locals and guests. Another important consideration needs to be made in creating new local jobs in order to promote the local economy. These activities should focus on the indigenous, authentic skills and knowledge of the residents of tourism regions, thus achieving independence and resilience for the tourism sector in Italy in a global context. According to the aforementioned strategic considerations, a portfolio of new resilient tourism products should still represent the history and identity of authentic Italian destinations. This aspect can become the key element in the nation's brand image. Within the current contribution, a resilient product is suggested under the name 'Make it and taste it'. Within this product proposal, gastronomy and wine tourism in Italy is further highlighted because this tourism theme represents an opportunity to revitalize and diversify tourism, promoting local economic development for new experimental destinations, involving many different professional sectors in alternative tourism regions.

This product orientation offers, in fact, many opportunities to increase and diversify professional skills and competences in experimental destinations in Italy, training new generations of workers whose main aim will be to enhance their native region and community. Reviving local development and local resources is the mission in this context. In order to succeed in this sense, there should be a perfect equilibrium and integration between material resources in the areas which characterize the tourism territories. In this way, the conservation of typical and traditional places will be achieved.

### 8.5.3 Ireland

According to the presented comparative analysis of sociocultural resilience, Ireland appears to have an admirable quality in reacting positively to the challenges presented by the current Covid-19 crisis and even far beyond that: the ability to preserve a destination's authenticity as well as all the operative functions to ensure basic supply and safety in a possible situation of a systemic crisis.

As an indigenous industry which has demonstrated a continued ability to grow, tourism's role in the Irish economy is uniquely significant. However, for tourism to keep contributing to Ireland's economic wellbeing, resilient strategies must focus on the needs of future consumers and the actions of international competitors, as well as being in line with the post-Covid-19 dynamics of the tourism market. A framework for the long-term resilient development of the tourism sector in Ireland is above all focused on its authentic and distinctive cultural heritage and rich local resources as important tools to highlight the resilient potential of new experimental tourism destinations in the country. Unique 'Irish' but locally-based touristic experiences include: awareness of alternative ways of life; distinctive European traditions; reconnection and revaluation of some universal human values like family devotion; and adherence to simpler ways of living.

In addition to its people, Ireland's other vitally important tourist attraction is the landscape. The scenic pastures and coastlines, and relatively good air and water quality, are a significant draw for visitors. The varied landscape across the country allows for the design of a diverse product portfolio on a nature tourism theme, with a wide range of diverse and complementary activities. Attention is therefore geared towards the involvement and creation of new tourism products which act as important components for the development of new niches in tourism, leisure and recreation. According to the comparative analysis of sociocultural resilience, Ireland provides a very limited number of competitive tourism destinations, especially in terms of diversity of tourism product portfolio and the level of tourism infrastructure. It can be seen from the study that the country has a very small number of World Heritage Sites and local/regional museums. These important elements should be developed and differentiated, offering several products to choose from. Investment is necessary for resilient product design. For instance, the focus should be on providing high-quality services and integrating existing products towards the extended tourism value chain to achieve a stimulating and rewarding product for target consumers. Following on from this, due to the small number of cultural and sport associations, tourism in Ireland should address opportunities for social interaction – for instance, developing the event sector and increasing the dialogical space between guests and locals. Furthermore, due to the high number of participants in church activities in Ireland, tourism could expand in relation to religion, so as to enhance in this sense the mental, physical and psychological strength of both locals and tourists thanks to religious tourism activities, such as pilgrim walks. In this way, consumers would be able to reduce stress and increase relaxation, serenity and personal resilience.

According to the strategic considerations made for Ireland, tourism should grow as an important economic activity in this country. It is especially relevant for the large number of coastal regions and offshore inhabited islands along the coast. The potential for coastal and marine tourism in Ireland to develop a new niche marine environment and to design related tourism products is substantial, especially given Ireland's world-renowned landscapes along 7700 km of coastline (Hynes *et al.*, 2019). This is why the resilient-oriented touristic products which perfectly fit into the Irish original habitat and atmosphere of yet unexplored potential destinations is joined under the theme 'The lighthouse trail' (with the image of lighthouses on the wild Atlantic way). This initiative in the product design highlights and protects the history, tradition and heritage of the island. Tourism and leisure in coastal and marine areas of Ireland comprise a significant component of the overall tourism market, which is particularly predestined to develop within innovative concepts of destination resilience. For this purpose, a portfolio of resilient products, which implement the safeguarding function for keeping alive the authenticity of Irish culture and land, is of particular importance. Furthermore, future developments of marine and coastal tourism should continue within the paradigm of destination resilience, particularly reflecting the highest standards of environmental quality and sustainability. The combination of these natural factors, indigenous health knowledge, inherited wisdom and simplicity of life, together with the authentic traits of a country, provide a source for the diversification and identity of resilient tourism products.

## 8.6 Conclusion

The empirical research and the comparative analyses on the sociocultural dimensions of destination resilience carried out for Germany, Italy and Ireland critically discusses and evaluates the strengths and weaknesses of potential experimental destinations in these countries, focusing on resilience as a driving concept for strategic rethinking and resetting of existing practices and measures. The indicators analysed and the indexes suggested for the current analysis of sociocultural resilience of the three countries have contributed to identify elements of tourism competitiveness, attractiveness and overall sustainability for the selected countries and, in particular, experimental destinations there.

The research shows that Germany and Ireland have the potential to overcome crises and react positively when critical situations occur through their functioning under the guiding paradigm of resilience. However, according to the results coming from the empirical research, Germany should focus on providing more wellness- and spiritual-oriented tourism products, as well as increasing education and training in the tourism sector, preserving sustainability and protecting its rich local culture. Ireland should focus on cultural heritage, propose new tourism products and increase opportunities for social interactions. Italy appears to be more vulnerable to shock or stress experienced at a given time, particularly to the current impacts of Covid-19. Italy has the potential to preserve culture and tradition, ethics and values but essential components of personal, community and sociocultural resilience should be better developed. From this perspective, Italy should focus on creating new local jobs, finding solutions to raise financial resources in touristic communities, increasing connectivity between locals and guests, and preserving culture, traditions and spirituality.

The outcomes of the research provide a solid basis for the implementation of resilience practices which deal with the increase of tourism attractiveness of the experimental resilient destinations in the selected countries and the competitiveness of their product portfolio. Thanks to the innovative resilient-oriented tourism products suggested in the current contribution, an innovative approach for rediscovering and restoring the historical and natural identity of touristic regions in the three countries can be realized. Furthermore, the uniqueness of the product portfolio of a destination is achieved by discovering a stronger sociocultural identity for the region, as well as providing local strengths and resources for tourism development. All these functional elements comprise a destination's potential resilient orientation. The comparative analysis of sociocultural resilience in Germany, Italy and Ireland shows that socially- and culturally-oriented cooperation provides the opportunity to initiate favourable shifts in tourism demand and to develop resilient business structures in even initially rather weak touristic regions. The continuation of scientific work in this field is of great importance. While it is a challenging perspective to bring to practice, the advantages in the long run can make the difference between destinations that are resilient to the ongoing dynamisms of society and those that are not.

However, the diversity of destinations represents the limitations of the universal use of the Code-Red Model for design of multi-level tourism products. In a further research step, it seems sensible to cluster destinations on the basis of different criteria and particular characteristics of sociocultural resilience. Among such criteria, resilience potentials, existing tourism infrastructures and thematic potentials for product design, as well as the degree of self-sufficiency could be mentioned. A categorization of destinations would modify the application of the Code-Red Model and enable further development aspects. However, the continuation of scientific work in this field is of great importance. The resilience perspective provides an interesting and promising avenue to further enhance destination design, development and management.

For instance, further considerations could also be made in terms of how stakeholders and residents cooperate in terms of initiatives that contribute to achieving the sociocultural components of destination resilience. Focused orientation and related implementation mechanisms are required to recognize the local social and cultural competences needed to build on the sociocultural resilience of diverse destinations. This also encourages local service providers to recognize the challenges and potentials which are specific to the sociocultural uniqueness of the countries and destinations, as well as to cooperate actively at the level of the destination to build on its socio-cultural resilience. More research is needed regarding the themes of building resilience and adaptive capacity-building in order to understand the resourcefulness of actors in the tourism industry in their quest to build more resilient tourism regions and learn from their successes and failures – to reflexively learn and adapt tourism literature to further understanding.

## References

Beeton, S. (2006) *Community development through tourism*. CSIRO Publishing, Clayton.

Calgaro, E.L., Dominey-Howes, D. and Lloyd, K. (2013) Application of the Destination Sustainability Framework to explore the drivers of vulnerability and resilience in Thailand following the 2004 Indian Ocean Tsunami. *Journal of Sustainable Tourism* 22(3), 361–383. DOI: 10.1080/09669582.2013.826231.

Cochrane, J. (2010) The sphere of tourism resilience. *Tourism Recreation Research* 35(2), 173–185. DOI: 10.1080/02508281.2010.11081632.

Cohen, G. and Bodeker, M. (2009) *Understanding the global spa industry: Spa management*. Butterworth Heinemann, Oxford.

Craft Industry Alliance (2019) An overview of the German craft market over the last 5 years. Available at: https://craftindustryalliance.org/an-overview-of-the-german-craft-market-over-the-last-5-years/ (accessed 15 October 2020).

Craftsman Project (2010) Report on the situation of craftsmanship in Europe. Available at: http://projects.ifes.es/pdfs/craft/craft1.pdf (accessed 13 October 2020).

Dahles, H. and Susilowati, T.P. (2015) Business resilience in times of growth and crisis. *Annals of Tourism Research* 51(2–4), 34–50. DOI: 10.1016/j.annals.2015.01.002.

Davoudi, S. and Porter, L. (2012) Resilience: a bridging concept or a dead end? *Planning Theory and Practice* 13(2), 299–333.

Duit, A., Galaz, V., Eckerberg, K. and Ebbesson, J. (2010) Governance, complexity, and resilience. *Global Environmental Change* 20(3), 363–368. DOI: 10.1016/j.gloenvcha.2010.04.006.

European Group on Museum Statistics (2020) National reports and publications. Available at: https://www.egmus.eu/ (accessed 13 October 2020).

Eurostat (2006) Participation in activities of churches or other religious organizations by level of activity limitation, sex and age. Available at: https://ec.europa.eu/eurostat/web/products-datasets/-/hlth_ds040 (accessed 15 October 2020).

Eurostat (2017a) Families with children in the EU. Available at: https://ec.europa.eu/eurostat/en/web/products-eurostat-news/-/EDN-20170531-1 (accessed 17 October 2020).

Eurostat (2017b) How many Europeans went to a cultural or sports event? Available at: https://ec.europa.eu/eurostat/en/web/products-eurostat-news/-/DDN-20170711-1 (accessed 20 October 2020).

Eurostat (2019) Employment rates of young people not in education and training by sex, educational attainment level and years since completion of highest level of education. Available at: https://ec.europa.eu/eurostat/web/products-datasets/-/edat_lfse_24 (accessed 20 October 2020).

Eurostat (2020a) Crude marriage rate and crude divorce rate. Available at: https://ec.europa.eu/eurostat/web/products-datasets/-/tps00206 (accessed 15 October 2020).

Eurostat (2020b) Tertiary education graduates: what are their subject? Available at: https://ec.europa.eu/eurostat/en/web/products-eurostat-news/-/EDN-20200124-1 (accessed 20 October 2020).

Federation of Irish sport (2017) Irish sport does matter. Available at: https://www.irishsport.ie/wp-content/uploads/2017/10/Sport-Matters-2017.pdf (accessed 13 November 2020).

Fontanari, M. and Ewald, T. (2018) The market of the tour operators in Germany within a disruptive environment: Implications for the management and the resilient orientation of destinations. In: Innerhofer, E., Fontanari, M., and Pechlaner, H. (eds) *Destination Resilience: Challenges and Opportunities for Destination Management and Governance*. Routledge (Taylor & Francis Group), Oxford, UK, pp. 184–201.

Fontanari, M., Traskevich, A. and Seraphin, H. (2021) (De)growth imperative: The importance of destination resilience in the context of overtourism. In: Andriotis, K. (ed.) *Issues and Cases of Degrowth in Tourism*. CAB International, Wallingford, UK, pp. 22–41.

German Olympic Sports Confederation (2018) Support the implementation of good governance in sport. Available at: https://www.siggs.eu/partner/dosb (accessed 10 November 2020).

Gössling, S., Scott, D. and Hall, C.M. (2021) Pandemics, tourism and global change: A rapid assessment of COVID-19. *Journal of Sustainable Tourism* 29(1), 1–20. DOI: 10.1080/09669582.2020.1758708.

Hall, C.M., Prayag, G. and Amore, A. (eds) (2018) *Tourism and Resilience: Individual, organisation and destination perspectives*, Series: Tourism Essentials: 5. Channel View Publications, Bristol.

Haller, M. and Yuchtman-Yaar, E. (2009) *International social survey program: Social Inequality I-IV ADD ON – ISSP 1987–1992–1999–2009*. Leibniz-Institut für Sozialwissenschaften, Gesis.

Helliwell, J.F., Layard, R. and Sachs, J.D. (2019) *World Happiness Report*. Available at: https://s3.amazonaws.com/happiness-report/2019/WHR19.pdf (accessed 20 October 2020).

Holling, C.S. (1996) Engineering resilience versus ecological resilience. In: Schulze, P. (ed) *Engineering Within Ecological Constraints*. The National Academies Press, Washington, DC, pp. 31–44.

Hynes, S., Aymelek, M., Corless, R. and Evers, N. (2019) A survey of marine and coastal overseas tourism activity in Ireland. Available at: https://www.nuigalway.ie/media/researchsites/semru/files/SEMRU-Overseas-Marine-Tourism-Report-Final1.pdf (accessed 20 November 2020).

Indecon International Economic Consultant (2010) Economic significance and potential of the crafts sector in Ireland. Available at: http://www.ccoienterprise.ie/downloads/Economic_Significance_and_Potential_of_the_Crafts_Sector_in_Ireland_-_Executive_Summary.pdf (accessed 17 October 2020).

Innerhofer, E., Fontanari, M. and Pechlaner, H. (2018) *Destination Resilience: Challenges and Opportunities for Destination Management and Governance*. Routledge (Taylor & Francis Group), Oxford.

Kauffman, D.L. (1980) *Systems One: An Introduction to Systems Thinking*. Future Systems Inc, Minneapolis, MN.

Lew, A.A. (2014) Scale, change and resilience in community tourism planning. *Tourism Geographies* 16(1), 14–22. DOI: 10.1080/14616688.2013.864325.

Luthe, T. and Wyss, R. (2014) Assessing and planning resilience in tourism. *Tourism Management* 44(3), 161–163. DOI: 10.1016/j.tourman.2014.03.011.

Mansfeld, Y. (1999) Cycles of war, terror, and peace: Determinants and management of crisis and recovery of the Israeli tourism industry. *Journal of Travel Research* 38(1), 30–36. DOI: 10.1177/004728759903800107.

Partale, A. and Partale, K. (2019) *Destination management in developing and emerging countries: Handbook and guidelines for building sustainable destination management organizations*. Deutsche Gesellschaft für Internationale Zusammenarbeit (GIZ).

Partale, A. and Partale, K. (2020) *Die Wertschöpfungskette Tourismus: Analyse und Anwendungsansätze für Projekte der Entwicklungszusammenarbeit* [The tourism value chain: Analysis and application approaches for development cooperation projects]. Deutsche Gesellschaft für Internationale Zusammenarbeit (GIZ).

Partale, K. and Traskevich, A. (2018) Methodological approach to the tourism development strategy and its application to Naroch destination. *Belarusian Economic Journal* 20(1), 131–146.

Pechlaner, H., Zacher, D., Eckert, C. and Petersik, L. (2019) Joint responsibility and understanding of resilience from a DMO perspective – an analysis of different situations in Bavarian tourism destinations. *International Journal of Tourism Cities* 5(2), 146–168. DOI: 10.1108/IJTC-12-2017-0093.

Prieto, M. (2020) *Design thinking for tourism product development: Application of Design Thinking techniques to achieve sustainable and memorable tourism experiences*. Sciencia Scripts.

Richter, L.K. and Waugh, W.L. (1986) Terrorism and tourism as logical companions. *Tourism Management* 7(4), 230–238. DOI: 10.1016/0261-5177(86)90033-6.

Ritchie, B.W. and Jiang, Y. (2019) A review of research on tourism risk, crisis and disaster management: Launching the annals of tourism research curated collection on tourism risk, crisis and disaster management. *Annals of Tourism Research* 79(1), 102812. DOI: 10.1016/j.annals.2019.102812.

Romão, J. (2020) Tourism, smart specialisation, growth, and resilience. *Annals of Tourism Research* 84, 102995. DOI: 10.1016/j.annals.2020.102995.

Sheppard, V.A. and Williams, P.W. (2016) Factors that strengthen tourism resort resilience. *Journal of Hospitality and Tourism Management* 28(3), 20–30. DOI: 10.1016/j.jhtm.2016.04.006.

Stienmetz, J., Kim, J., Xiang, Z. and Fesenmaier, D.R. (2020) Managing the structure of tourism experiences: Foundations for tourism design. *Journal of Destination Marketing & Management* 100408.

Swidler, A. (1986) Culture in action: Symbols and strategies. *American Sociological Review* 51(2), 273–286. DOI: 10.2307/2095521.

Tomej, K. and Xiang, Z. (2020) Affordances for tourism service design. *Annals of Tourism Research* 85, 103029. DOI: 10.1016/j.annals.2020.103029.

Traskevich, A. and Danskikh, S. (2018) Cooperative approaches between health- and agro-tourism stakeholders in increasing rural destination resilience in Belarus. In: Innerhofer, E., Fontanari, M., and Pechlaner, H. (eds) *Destination Resilience: Challenges and Opportunities for Destination Management and Governance*. Routledge (Taylor & Francis Group), Oxford, pp. 77–87.

Traskevich, A. and Fontanari, M. (2018) Mental wellness in resilient destinations. *International Journal of Spa and Wellness Industry* 1(3), 193–217. DOI: 10.1080/24721735.2019.1596656.

Traskevich, A. and Fontanari, M. (2019) *Fuzzy-clustering approach for strategic development of resilient destinations for mental wellness*. Tourism into the new decade: challenges and prospects: Book of abstracts of the International Conference on Tourism (ICOT2019), Braga, Porto, Portugal, 26–29 June 2019, pp. 105–106.

Traskevich, A. and Fontanari, M. (2021) Tourism Potentials in Post-COVID19: The Concept of Destination Resilience for Advanced Sustainable Management in Tourism. *Tourism Planning & Development* 1–25. DOI: 10.1080/21568316.2021.1894599.

Uğur, L. (2018) Integrated destination governance: an evolutionary approach to open innovation and stakeholder engagement for resilient tourism destinations. In: Innerhofer, E., Fontanari, M. and Pechlaner, H. (eds) *Destination Resilience: Challenges and Opportunities for Destination Management and Governance*. Routledge, Oxford, pp. 88–98.

Volgger, M., Erschbamer, G. and Pechlaner, H. (2021) Destination design: New perspectives for tourism destination development. *Journal of Destination Marketing & Management* 19(1), 100561. DOI: 10.1016/j.jdmm.2021.100561.

Volo, S. (2021) The experience of emotion: Directions for tourism design. *Annals of Tourism Research* 86(1), 103097. DOI: 10.1016/j.annals.2020.103097.

Walder, B., Weiermair, K. and Sancho Pérez, A. (2005) *Innovation and Product Development in Tourism*. Erich Schmidt Verlag, Berlin.

# 9 Planning for a More Sustainable Tourism? A Pan-Nordic Analysis of Regional Tourism Strategies for Rural Areas

**Rikke Brandt Broegaard[1]*, Ágúst Bogason[2] and Anna Karlsdóttir[2]**
*[1]Centre for Regional and Tourism Research, Bornholm, Denmark; [2]Nordregio, Skeppsholmen, Stockholm, Sweden*

## 9.1 Introduction

Sustainability is highlighted as a central issue in the national tourism policies of the Nordic countries (see e.g. Øian *et al.*, 2018, for an overview), and tourism development plans (TDPs) provide an interesting source for analysing the degree to which sustainability concerns are integrated into the tourism sector at a subnational level. This chapter explores visions and goals for sustainable tourism development through a content analysis of 110 rural municipal, subregional and regional tourism plans from the Nordic countries – Denmark, Finland, Iceland, Norway, Sweden and the Faroe Islands, Greenland and Åland. The TDP content mapping is especially interesting, since these documents are formulated at the lowest administrative level(s) by tourism-related actors or planners at municipal, subregional and/or regional level, and with a few exceptions are available throughout the Nordic territory. It is expected that with growing public awareness about the need for transition to more sustainable tourism practices, along with political statements and commitments to do so, sustainability concerns will be strongly reflected in subnational tourism development plans.

Increased interest from the Nordic Council of Ministers to learn from neighbouring Nordic countries' tourism development spurred a research project analysing tourism planning in the Nordic countries, with special emphasis on sustainable development. The project also mapped other issues that are given attention in the subnational, rural tourism development plans (Bogason *et al.*, 2020). This chapter is built on the project report, with special focus on the reflection of sustainability concerns in the TDPs.

The chapter first presents the theoretical background of the study, followed by a presentation of the methods used for the identification of subnational TDPs from rural areas across the Nordic territory, and their coding and sustainability concern typology is explained. It then presents results for the different ways in which sustainability concerns are reflected in the analysed material. It explores two different patterns to understand the resulting distribution of sustainability concern types: (i) the local openness about negative impacts

*Corresponding author: rikke.brandt.broegaard@crt.dk

 *Tourism Planning and Development in Western Europe* (eds K. Andriotis *et al.*)
DOI: 10.1079/9781800620797.0009

of tourism, and (ii) (a proxy for) the degree of participation in the development of the subnational TDP relative to the resulting sustainability concern. The chapter ends with a short discussion and conclusions.

## 9.2 Theoretical Background

Recent academic literature on tourism has emphasized a paradigm shift that highlights a new central role for tourism in local development, and in the regeneration of both regions and destinations (Luthe and Wyss, 2014; Amore *et al.*, 2018; Innerhofer *et al.*, 2018). A key element in this shift is the need for tourism development to be planned and managed so that the original character and authenticity of destinations are preserved for future generations (Bramwell and Lane, 2010; Costa *et al.*, 2018). For this, it is also necessary to view tourism more broadly, as an integrated 'part of the larger socio-economic bio-physical system' (Hall, 2009, p. 53). In Nordic settings, tourism is increasingly seen as a source of place development, innovation and transformation, especially in rural or peripheral settings where tourism growth raises great expectations for local renewal and economic prosperity (Müller, 2013; Broegaard, 2020). The importance of tourism-generated employment may be especially pronounced in more sparsely populated rural areas, where structural change has resulted in a decline in traditional job-creating rural sectors like farming, forestry and fishing (Müller and Jansson, 2007; Øian *et al.*, 2018). Until the onset of the Covid-19 pandemic, increasing mobility and tourism growth made tourism reach even peripheral rural communities, which until a few years ago had next to no tourists.

Literature on sustainable tourism simultaneously emphasizes governance and management as key factors for such a transition (Hall, 2019). Several studies highlight the role of subnational governments and institutions in leading this transition by stressing the importance of broad participation in democratic planning processes (Bramwell, 2011; Bramwell and Lane, 2011; Ruhanen, 2013), and some studies have highlighted the need for greater openness about both the positive and negative impacts of tourism (Kauppila *et al.*, 2009; Higgins-Desbiolles *et al.*, 2019). Tourism development plans form an appropriate material for analysing the degree of openness about diverse tourism impacts.

Considering global climate change and sustainability challenges, it becomes relevant to examine whether, and how, subnational tourism plans for rural areas take sustainability into consideration. The chapter asks the following questions: To what extent do local and regional actors use tourism development plans as tools to balance economic development and environmental or social pressures from tourism? Does openness about positive as well as negative impacts of tourism influence the way in which sustainability concerns are handled in local tourism development plans? Does the level of participation in the process leading up to a tourism development plan influence the handling of sustainability concerns?

The ambiguity of the concepts of sustainability and sustainable tourism is highlighted in several studies as something that detracts from the effectiveness of sustainability-related efforts, for example regarding the implementation of sustainable tourism indicators (Torres-Delgado and Saarinen, 2014), or the tourism sector's absolute social and environmental impact (Hall, 2011). The analysed material also shows considerable ambiguity regarding how the concept of 'sustainability' was used. The chapter presents a typology developed to reflect the varied ways in which 'sustainability concerns' were operationalized in the analysed material, to handle – and map – this ambiguity.

## 9.3 Studying Sustainability Concerns in Subnational Tourism Plans in the Nordic Countries

### 9.3.1 Methods and context

A web search was carried out to identify and retrieve all recent available subnational destination management plans (DMPs) from Nordic rural regions (August–October 2018, with additional searches for Finnish material February–March 2019). Search terms included destination management plan, tourism policy, tourism strategy, tourism plan, local, regional, etc., and the search was carried out both in English and in the primary language for each country. Additional searches were carried out for areas with missing results. To make the geographical coverage of the documents more complete, subnational plans and documents explicitly addressing tourism planning were included in the screening, even if they were not DMPs per se, but for example regional economic development plans with a section devoted to tourism development. The choice to include plans addressing tourism development at the subnational level, even if these plans were drawn up by actors other than destination management organizations (DMOs, also known as destination marketing organizations), results in a relatively broad range of authors of the analysed plans, including DMOs, municipal governments (as well as consortia of multiple neighbouring municipal governments); regional development authorities; regional business development entities; growth forums; and in some instances, platforms specifically created to develop a DMP or tourism strategy. Processes leading to the plans also differed considerably.

After screening for relevance regarding (i) rural area (type II–V in Nordregio's urban–rural gradient typology); (ii) valid or most recent report; and (iii) subnational area, i.e. municipalities, regions, 'landscapes' or tourism destinations for the five largest Nordic countries (for the Faroe Islands, Greenland and Åland, however, this last criterion had to be ignored, as only national-level plans existed), the resulting empirical material for the analysis consisted of 110 reports containing tourism development planning at subnational level for rural areas within the Nordic region (Fig. 9.1).

The collected TDPs were each carefully read and coded (using the Nvivo software) for content dealing with diverse aspects of tourism planning, including growth and sustainability concerns, as well as some factual and procedural elements. The content coding was carried out by a multinational, multilingual team, dividing the TDPs by nationality and language, taking measures to ensure a homogeneous coding. Several content-coded variables were predefined, following research questions and subquestions. Other variables, however, arose as being potentially important during the careful reading and coding process. Please see Bogason *et al.* (2020) for a full list of the variables used during the coding.

Relevant attributes were coded in classification sheets. For example, classes were constructed for the duration of the report's validity (short, medium or long-term, or unassigned; Table 9.1), or the administrative level for which the document was valid (municipal, subregional, regional or national, Table 9.2), or whether it mentioned any experiences with negative consequences of tourism (initially classified as not, somewhat, and much; later reduced to no or yes, Table 9.5). Other classifications targeted the TDP development process (Table 9.7), the sustainability concern type (Tables 9.3, 9.4), or whether cruise tourism was mentioned in the TDP, whether the TDP was from a border region, or whether the tourism described was predominantly nature or culture oriented, or a combination (tables not shown for these latter classifications). These attributes were recorded to describe the reports and to create an overview, but also to facilitate further analysis along the lines of these attributes. A detailed list of classifications may be found in Bogason *et al.*, 2020.

## 9.4 Subnational TDPs in the Nordic Countries

Thematically, most of the collected plans broadly addressed tourism planning issues within their geographic coverage. The reports differed considerably in content and focus due to local and regional differences in tourism potential and resources, and in priorities and approaches. A few of the analysed plans addressed a thematic focus, such as gastro tourism or cruise tourism, for the geographic area.

There is a large variation in the duration of the TDPs analysed, ranging between 2 and 15 years (Table 9.1). The medium-term duration (5 to 8) was most frequent (with 42 TDPs, or 38%), followed by a short-term lifespan (up to 4 years; a total of 34 TDPs, 31%), while less than a sixth (17 TDPs, 15%) of the sample had a long time frame (9 or more years), whereas no time frame was stated for the final 17 TDPs.

TDPs elaborated at the regional level is most common in the Nordic region, accounting for

**Table 9.1.** TDPs classified by duration.

| TDP duration (valid time period in years) | Number of TDPs |
|---|---|
| Short term (up to and including 4 years) | 34 |
| Medium term (5–8 years) | 42 |
| Long term (9+ years) | 17 |
| Unassigned | 17 |
| *Total* | *110* |

more than half of the sample, followed by TDPs elaborated at municipal level, which add to one third of the sample. TDPs at subregional level, including for example TDPs elaborated collaboratively between some neighbouring municipalities without covering an entire region, are less common, with only 12 of the 110 cases. Some areas are covered by more than one TDP (where both municipal and regional or subregional TPDs exist). As mentioned in the methodology section, only national level documents were identified for the Faroe Islands, Greenland and Åland. The table below summarizes the distribution of the subnational TDPs from each Nordic country by geographical or administrative level coverage (Table 9.2).

Relative to the overall sample, Denmark stands out as having more TDPs elaborated at municipal level (a total of 60% of the Danish subnational TDPs), whereas Iceland and Norway stand out as having more TDPs elaborated at regional level (86% and 82% of the Icelandic and Norwegian analysed subnational TDPs, respectively). The collected reports include four TDPs at national level, from the three autonomous regions of Åland, the Faroe Islands and Greenland.

The map (Fig. 9.1) shows the geographical coverage of the gathered material. As indicated by the legend, some areas are represented in both municipal or subregional and regional TDPs. The grey areas indicate either urban regions, or areas for which no TDP was identified, despite considerable effort being put into the data identification at the beginning of the research project (see Bogason *et al.*, 2020, for a complete list of the reports analysed).

Our TDP analyses indicate that economic growth is generally viewed as the most sought-after benefit of tourism. The focus in most of the analysed TDPs is on the numbers of visitors (possibly segmented into different types of tourist), the number of overnight stays and the locally created jobs. Generally, the stated goals were to increase numbers for these three parameters, and often with an ambition for high growth rates (of between 5 and 10% per annum increase in tourism economy), with some extreme growth ambitions over 10% annually. Only in a few cases were growth ambitions formulated more sophisticatedly to include goals like increasing the local value added per tourist or achieving greater economic income from tourism experiences. Local job creation expectations were often mentioned, but seldomly quantified. Few TDPs did not mention job creation explicitly, as tourism is seen as a source of job creation, with a possible multiplier effect providing a boost in revenues for a range of other industries. The

**Table 9.2.** Distribution of TDPs in each Nordic country, by administrative level or coverage.

| TDP coverage | Municipal | Subregional | Regional | National | *Total* |
|---|---|---|---|---|---|
| Denmark | 18 | 6 | 6 | 0 | *30* |
| Finland | 6 | 1 | 14 | 0 | *21* |
| Iceland | 1 | 0 | 6 | 0 | *7* |
| Norway | 2 | 2 | 18 | 0 | *22* |
| Sweden | 9 | 3 | 14 | 0 | *26* |
| Faroe Islands | 0 | 0 | 0 | 1 | *1* |
| Greenland | 0 | 0 | 0 | 2 | *2* |
| Åland | 0 | 0 | 0 | 1 | *1* |
| *Total* | *36* | *12* | *58* | *4* | *110* |

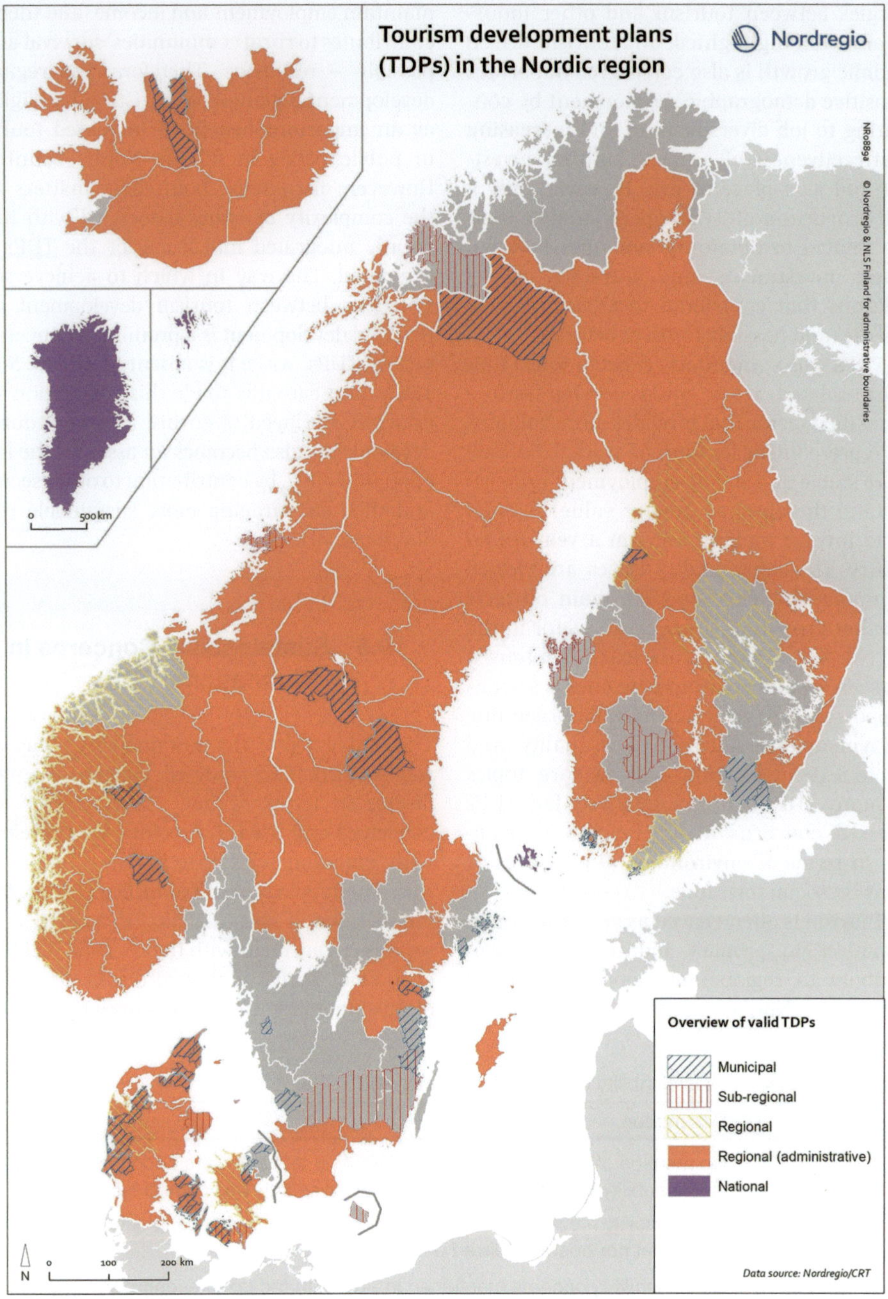


**Fig. 9.1.** Map showing the regions covered by the TDPs gathered for the research. (Elaborated by Nordregio cartographer Johanna Jokinen, based on input from the authors.)

Note: 'Regional (administrative)' plans (rose colour) were developed by a formal administrative region. 'Regional' plans (yellow, striped) were plans developed by two or more regions together, or between municipalities within different administrative regions.

synergies between tourism and other industries are often highlighted, and tourism-driven economic growth is also considered important for positive demographic development by contributing to job diversification and increasing the attractiveness of a place, both for investment and as a place to live. However, only a small proportion of TDPs operationalize their goals related to obtaining synergies between tourism investments and attractive living conditions that contribute to attracting new residents (and possibly tourism actors).

TDPs also mention expectations that tourism-related work could develop into a more skilled sector that provides more valuable jobs. A prerequisite for tourism work to become a more stable and steady employment proposition (and therefore of greater value) is often said to involve making tourism a year-round industry. Thus, seasonality issues are viewed as a major challenge and the main obstacle to further growth of the tourism sector in the rural Nordics. Consequently, extending tourist high seasons and creating new ones is seen as an opportunity to grow and to enable a genuine employment diversification. Seasonality, and how to extend existing seasons, are topics that many TDPs address. Relatively few TDPs expressed concerns about tourism development in terms of environmental pressures or strains on social tolerance.

Tourism is often viewed as one of the main alternatives to primary industries that can contribute to regional development in rural areas of the Nordic countries, as it helps to maintain employment and income, and thereby contributes to rural communities' survival and – possibly – resilience. Therefore, the regional development potential of tourism is highlighted as an important benefit of increased tourism in public policy in all the Nordic countries. However, discussion about these matters and the complexity of issues associated with them is only integrated into a few of the TDPs we examined. The way in which to achieve such synergies between tourism development and regional development is a prominent concern in several TDPs, while it is absent in others. Some TDPs have carefully made this connection and prioritize the need to ensure that any tourism development also becomes an asset for the local population (e.g. by contributing to other sectors, and thus encouraging more sustainable rural development).

## 9.5 Sustainability Concerns in Subnational TDPs

The ambiguity of the sustainability concept is well known. The analysed material was very diverse in terms of how 'sustainability' was conceived and used. A typology was developed with 'emerging' classes to handle – and map – this. The typology resulted in five classes, from 'no mention of sustainability' to a detailed and serious engagement with the concept, including economic, social and environmental aspects of sustainability. Table 9.3 defines the different

**Table 9.3.** Typology of sustainability concerns in the analysed TDPs.

| Type | | Description |
|---|---|---|
| Absent | | No mention of sustainability in the TDP (If no text bits were coded under 'sustainability concerns' an additional word search was carried out in Nvivo) |
| General | | Sustainability mentioned in the TDP, as reference to national or international goals, but not operationalized or integrated further |
| Economic focus | | Sustainability concerns mentioned in TDP with regard to economic sustainability or viability of tourism businesses |
| Holistic | Within tourism | Sustainability concerns mentioned and elaborated on, including economic, social and environmental aspects thereof, and some operationalization with respect to the tourism sector |
| | Broader | Sustainability concerns mentioned and elaborated on, including economic, social and environmental aspects thereof, and some operationalization with respect to the tourism sector and sectors outside thereof |

**Table 9.4.** Distribution of subnational TPDs by country by sustainability concern type.

| TDP sustainability concern type | Absent | General | Economic | Holistic | | *Total* |
|---|---|---|---|---|---|---|
| | | | | Within tourism | Broader | |
| Denmark | 11 | 8 | 3 | 2 | 6 | *30* |
| Finland | 3 | 10 | 0 | 2 | 6 | *21* |
| Iceland | 0 | 1 | 1 | 3 | 2 | *7* |
| Norway | 1 | 2 | 1 | 5 | 13 | *22* |
| Sweden | 1 | 1 | 8 | 10 | 6 | *26* |
| Faroe Islands | 0 | 0 | 0 | 0 | 1 | *1* |
| Greenland | 1 | 1 | 0 | 0 | 0 | *2* |
| Åland | 0 | 0 | 1 | 0 | 0 | *1* |
| *Total* | *17* | *23* | *14* | *22* | *34* | *110* |

sustainability concern types. Examples from TDPs for each type can be found in Bogason *et al.* (2020, pp. 61–62).

Each TDP was assigned into a type, depending on the degree to which it took sustainability concerns into consideration and the degree to which the treatment included economic, social and/or environmental aspects.

Table 9.4 shows considerable national differences regarding the way in which sustainability is treated in the analysed subnational TDPs. See also Fig. 9.2.

Fig. 9.2 shows a map of the distribution of different sustainability concern types in the TDPs in the Nordic territory. To represent the different geographical coverage of the TDPs analysed, the map is divided into regional and subregional levels. The subregional level mainly consists of single municipal plans, in addition to subregional plans and TDPs that cover (e.g.) coastal area or an archipelago that crosses over regional/municipal borders. The regional level of the map shows the sustainability concern types of the regional-level TDPs.

The distribution of TDP 'sustainability concern' types for the entire material showed that the TDPs fall in two halves. One half does not operationalize sustainability concerns, or only does so in general terms or with an economic focus within the tourism sector. The other half conceives sustainability as integrating economic, social and environmental elements, which require operationalization at the local, subregional or regional level. The diagonally shaded areas in Fig. 9.2 represent those TDPs that make no mention of sustainability. Considering the fundamental importance of sustainability issues globally, societally and as a sector, it was surprising that as many as 15.5% of the TDPs did not mention 'sustainability' in any way. The dark grey areas represent TDPs who mention the concept of sustainability, in a limited way, as a reference to an overall, global or international or national agenda, but without further operationalizing or setting any goals regarding this for the local context. TDPs in this category make up 20.9% of the total TDPs. Together, these two types represent more than a third of the TDPs. The brown areas represent those TDPs which limit their discussion of sustainability to the economic sustainability of tourism businesses. This type of TDP accounts for 12.7% of the TDPs. Thus, 49% of the TPDs, those in these three categories, show very limited interest in or operationalization of sustainability concerns.

The green areas, which together represent about half of the TDPs (51%), discuss sustainability concerns in a way that integrates economic, social and environmental concerns, and with some degree of local operationalization and engagement. We have labelled this a 'holistic' sustainability approach. Within the green category, the light green TDPs make up about a third (or 20% of the total sample), with a tourism-centred sustainability treatment. They come predominantly from Sweden, as well as from Iceland and Norway. The dark green areas, which represent about two-thirds of the green category (or 30.9% of the total sample) represent TDPs with a holistic sustainability

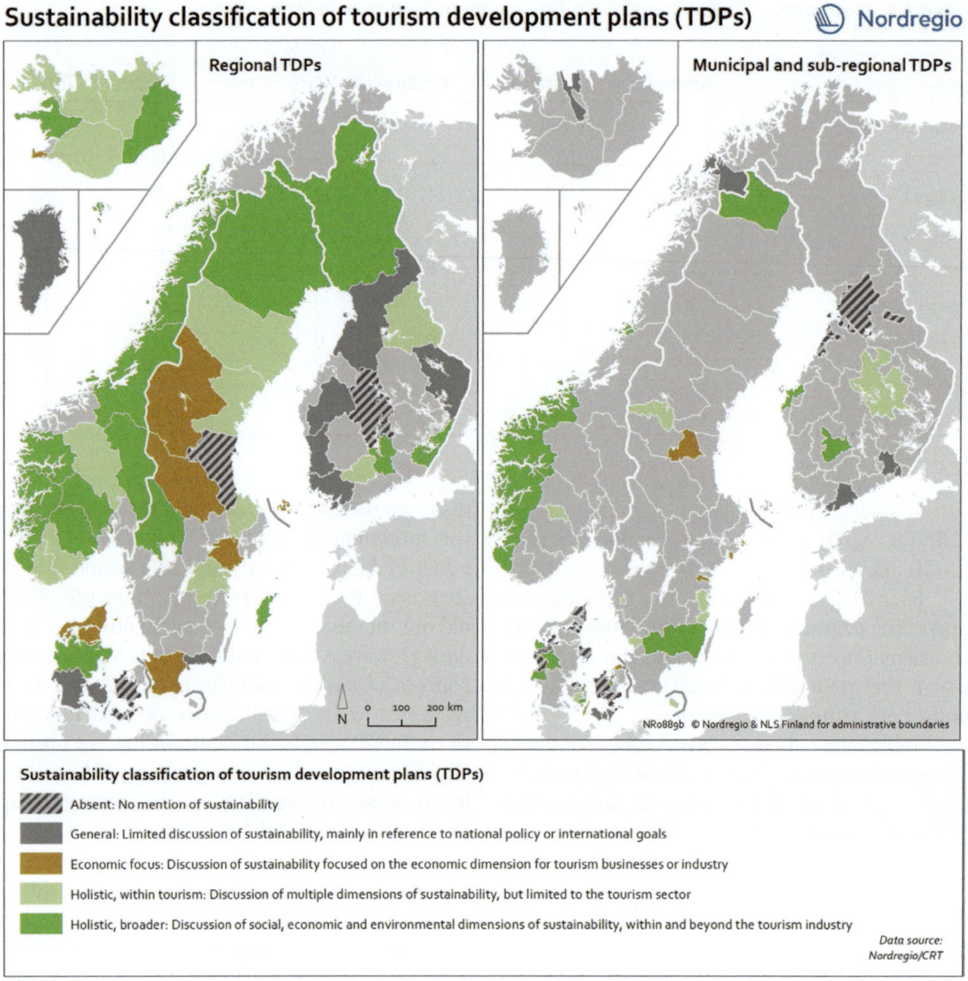


**Fig. 9.2.** Geographical distribution of TDPs according to their assigned sustainability classification type. (Elaborated by Nordregio cartographer Johanna Jokinen, based on input from the authors.)

treatment that also includes wider societal effects and considerations in their discussions of sustainability. They predominantly come from Norway, followed by Sweden and Finland.

National differences are likely to reflect a combination of different tourism type characteristics in each country and its regions, the associated pressure on natural and social resources, as well as the degree to which national authorities and national tourism actors have engaged with sustainable tourism transformations. While sustainable tourism development has a prominent place in all Nordic national tourism development frameworks, Norway stands out as having set ambitious national goals, in a way that is frequently reflected in the subnational TDPs, and where concrete, local ambitions are often framed with reference to national guidelines. As early as 2007, the national tourism strategy placed the development of Norway as a sustainable destination as a precondition for the government's business policy regarding the tourism sector, as well as to national initiatives regarding tourism innovation. It is our general impression from coding the material that the Norwegian subnational TDPs refer highly positively to national strategies.

The TDPs identified for the Nordic rural regions represent a broad range of document types, constituted from different types of actor, through different types of process, and for varying purposes. This variety in type and purpose can explain some of the observed variation. However, considering this diversity, some patterns still stand out. Of 30 and 21 TDPs analysed for Denmark and Finland, respectively, sustainability concerns are mostly classified as being 'absent' or 'general' (19 of 30 in Denmark, and 13 of 21 in Finland), meaning that the TDPs either do not mention sustainability concerns, or that they are addressed in a way that mentions this agenda at national and/or international level, without operationalizing it at the subnational level. The first type is the single most frequent type in Denmark, whereas the latter is the single most frequent type for Finland. Sweden, Iceland and Norway have most of their TDPs classified as light or dark green, meaning that they treat sustainability concerns in a holistic way, either by focusing on the tourism sector or by including elements of relevance to wider society (16 of 26 in Sweden, 5 of 7 in Iceland and 18 of 22 in Norway). Of these, Norway stands out as being the country where the dark green TDP classification is the single most frequent type.

However, even though a TDP treats economic, social and environmental sustainability concerns holistically, and does so in an operationalized way, internal contradictions may exist in the TDPs. For example, several TDPs discuss sustainability concerns in detail and set goals within this agenda, while simultaneously aiming to increase the number of visitors, including those who arrive by air, or the number of cruise tourists, despite the known environmental costs of these tourism types. Such internal contradictions are not reflected in the typology.

### 9.5.1 Do negative experiences affect sustainability concerns?

Media debates and academic literature increasingly reflect that high numbers of tourists put pressure on fragile environments, public goods (including regional and local infrastructure, and even the long-established Nordic tradition of the public right of access) and social wellbeing. However, in the TDPs, such negative experiences are hardly reflected, neither as local experiences, from which local lessons can be learned and similar experiences avoided in the future, nor as lessons learned from other destinations. Most likely, there is an 'underreporting' of negative experiences from tourism in the TDPs, many of which present tourism in as positive a light as possible. Overall, negative experiences are only addressed in 13 out of the 110 analysed TDPs (12%), as shown in Table 9.5.

**Table 9.5.** Distribution of subnational TPDs by country by mention of negative tourism experiences.

| Mention of negative tourism experience (local or as lesson learned from elsewhere) | No | Yes | *Total* |
|---|---|---|---|
| Denmark | 30 | 0 | *30* |
| Finland | 19 | 2 | *21* |
| Iceland | 2 | 5 | *7* |
| Norway | 17 | 5 | *22* |
| Sweden | 26 | 0 | *26* |
| Faroe Islands | 0 | 1 | *1* |
| Greenland | 2 | 0 | *2* |
| Åland | 1 | 0 | *1* |
| *Total* | *97* | *13* | *110* |

There is a clear geographical pattern in the distribution of reporting of negative experiences, and none of the subnational TDPs from Denmark or Sweden mention negative tourism effects. The tourism plan of the Faroe Islands discusses negative impacts in some detail, as do some of the Finnish and Norwegian subnational TDPs, while most of Iceland's subnational TPDs mention such impacts. The geographical distribution suggests that negative experiences from tourism pressures (including cruise tourism) are more likely to be reported when pristine areas are involved.

By cross tabulating sustainability concern type and whether experiences of negative effects from tourism are mentioned in TDPs, a pattern is identified: recognition of negative consequences of tourism is associated with TDPs that have a holistic treatment of sustainability concerns. In fact, all the TDPs that mention

**Table 9.6.** Sustainability concern type by negative tourism experience mentioned in TDP ($n$ = 110). Percentage of TDPs according to mentioning negative experience, by sustainability concern.

| | | Sustainability concern type in TDP | | |
|---|---|---|---|---|
| TDP negative experience mentioned?*** | | Absent, general, or economic | Holistic (whether within tourism or beyond) | Total |
| No | ($n$ = 97) | 55.7 | 44.3 | 100 |
| Yes | ($n$ = 13) | 0.0 | 100.0 | 100 |

***$p$<0.001, Pearson chi-squared test of correlation.

some experiences of negative tourism effects fall into the 'green' categories of sustainability concern (Table 9.6, where the five sustainability concerns have been combined into two groups, due to the small number of TDPs which mention negative tourism experiences). The association is statistically significant at the 0.1% level, using a chi-squared test. Thus, those TDPs that have mentioned having or working to prevent negative effects of tourism are those that are most likely to treat sustainability concerns in a holistic matter. Actually, 10 of the 13 TDPs that mention negative tourism experiences fall within the 'holistic, broad' sustainability concern type, which discusses all three sustainability dimensions in ways that go beyond the tourism industry.

### 9.5.2 Association between TDP processes and sustainability concerns

This subsection examines associations or patterns between the process leading to the TDP and the treatment of sustainability concerns in the TDP. As briefly mentioned in the methodology section, a typology was developed for the distinct processes upon which the TDPs were elaborated, as the initial TDP reviews suggested distinct differences. The TDP process typology contained four classes, based on the degree of participation mentioned in the introduction or methods section of the TDP (if at all), and if no participation was mentioned (or no external actors thanked for their contributions or participation), the TDP was classified under the 'driving' actor. The process classes spanned from: (i) inclusive processes, in which a broad spectrum of local actors, in addition to the tourism actors (e.g. through hearings, public meetings, surveys, chat-forums, etc.) were involved and (ii) inclusive processes involving consultation with tourism actors, to (iii) consultancy-led processes (where the process was outsourced by the territorial actor to a company, and no participation was mentioned), and (iv) institutionally-led processes (e.g. by a municipality or a region, or a DMO), without mention of participation from other actors during the process. However, as only six cases were classified as 'consultancy-led', Table 9.7 combines institutionally- and consultancy-led processes into one group for the purpose of continuing with statistical tests. The three-category typology is a proxy for assessing the degree to which the TDP process was participatory, spanning from limited to broad participation. We have limited this analysis to focus on the subnational TDPs, i.e. excluding the (national-level) TDPs from the Faroe Islands, Greenland and Åland, as it cannot be expected that national-level processes show the same degree of local participation as regional, subregional and municipal ones.

Table 9.7 shows a cross-tabulation of the TDP process classifications according to their sustainability concern typology for the 106 analysed subnational TDPs. While the table indicates that any type of TDP process can lead to any type of sustainability concern in the resulting documents, the distribution in the table shows an association between the least participatory process (the institutionally-led and consultancy-led TDPs) and TDPs where sustainability concerns were lacking or not operationalized. Similarly, an association exists between the inclusive processes (whether limited to the tourism sector or extending to include a broad range of societal actors) and TDPs that have a holistic concept of sustainability concerns and treat these in some or much detail. Due to

**Table 9.7.** Sustainability concern type by TDP process. Percentage of TDPs according to process, by sustainability concern ($n$ = 106).

| TDP process*a | | Sustainability concern type in TDP: Absent | General | Economic | Holistic tourism | Holistic broad | Total |
|---|---|---|---|---|---|---|---|
| Institutionally- and consultancy-led | ($n$ = 30) | 30.0 | 16.7 | 20.0 | 20.0 | 13.3 | 100 |
| Inclusive, tourism actors | ($n$ = 46) | 8.7 | 15.2 | 10.9 | 28.3 | 37.0 | 100 |
| Inclusive, broad range actors | ($n$ = 30) | 10.0 | 33.3 | 6.7 | 10.0 | 40.0 | 100 |

*$p$<0.05, Pearson chi-squared test of correlation.
[a]4 cells (26.7%) have expected count less than 5. The minimum expected count is 3.68.

the distribution, with a relatively low number of observations (and expected count) in some cross-tabulation cells, we are just outside the border where chi-squared tests are permitted. With this reservation, however, the association between participatory processes and holistic sustainability concern treatment is statistically significant at a 5% level.

As the table shows, close to half of the TDPs developed in an institutionally- and consultancy-led process with limited public participation, either did not mention sustainability, or only in a very limited way. At the other end of the spectrum, half or more of the TDPs that were developed in an inclusive participatory process treat sustainability concerns in a holistic way. For those TDPs where the inclusive process was limited to tourism actors, as many as 65% of the TDPs deal with sustainability issues in a holistic way. For those TDPs where the inclusive process included a broad range of actors in addition to the tourism actors, 50% of the TDPs address sustainability concerns holistically, and as many as 40% of them treat sustainability not only in a holistic way, including economic, social and environmental elements, but also as extending beyond the tourism sector as such.

Based on the observed distribution, it can be hypothesized that moving away from a traditional business-development authority-led process, with limited outside participation, towards more of a community-based destination development process, with a broader collaboration of different sectors, including civil society (thereby allowing multiple viewpoints and experiences to be voiced in the process), leads to a more holistic end product. Those TDPs that were developed in a participatory manner with a broad range of local actors tended to take more factors into account when discussing sustainability concerns in tourism development, while those TDPs developed without broad participation tended to focus more, and sometimes exclusively, on economic aspects of sustainability, seen from the perspective of tourism businesses, i.e. without the inclusion of externalities.

Analysis of the level of administrative or geographic TDP coverage (still excluding the national-level TDPs) and sustainability concerns showed no statistically significant association. However, there is a slight tendency towards regional-level TDPs expressing sustainability concerns that are classified as 'holistic', but the trend is not strong enough to be statistically significant (and no table is shown). Likewise, no statistically significant association was found when analysing whether certain administrative or geographic levels were associated with more participatory processes. However, as with the above association, there is a tendency for the TDPs carried out at municipal level to more often have been elaborated through a participatory process involving tourism actors, while the TDPs carried out at regional level more often include a broader range of actors. The subregional TDPs tend to be institutionally- or consultancy-led processes, possibly because firmly established institutions or networks to lead such processes in more participatory ways are lacking at the subregional level, which is composed of multiple administrative units.

Furthermore, it was tested whether the time span for a TDP showed any association with sustainability concern type, based on a hypothesis that a longer time span might be associated with stronger sustainability concerns. However, no such association was identified in the analysed material (and no tables are shown for this analysis).

## 9.6 Discussion and Conclusion

This chapter has explored visions and goals for sustainable tourism development through a content analysis of 110 rural municipal, subregional and regional tourism plans from the Nordic countries in order to map sustainability concerns in recent plans at the lowest administrative level(s) by tourism-related actors or planners throughout the Nordic territory. At national and international levels, sustainability concerns are considered the greatest challenge to our societies today. Given that Nordic countries are considered frontrunners in sustainable development and green transition, and all national tourism strategies have sustainability as an overarching theme, one would assume that they would address sustainability concerns strongly in the subnational strategies too. However, our analysis shows that just over half of the TDPs take sustainability concerns into account in what we call a holistic way, by considering the social, environmental and economic dimensions. Of these 'green' TDPs, about a third limit their sustainability considerations to what the tourism sector controls, while the rest include sustainability aspects beyond the tourism sector. The other half of the TDPs does not take sustainability into account in a holistic way, as the concept is either absent, only treated in general terms or limited to economic sustainability within the tourism sector.

Following what Kauppila *et al.* (2009) would call the 'tourism as an industry' discourse, one could argue that sustainable development should not be the prime concern of DMOs and tourism actors who are focusing on the viability of their businesses. The business performance and growth-oriented aspects are well reflected in the subnational TDPs. Most TDPs set goals for economic growth within the tourism sector, either measured by increased number of overnight stays, increased turnover and/or increased indirect consumption created by tourism. However, if the job of the TDPs is to formulate a road map for local tourism actors to collectively position the tourism destination positively within a wider vision of tourism development, such a map must fully take on board the positive effects, challenges and requirements of sustainable development. Viewed in this light, it is surprising that TPDs could have been formulated without the slightest mention of the greatest challenge to societies in general, and to the tourism sector in particular.

While the data show considerable variation regarding the extent to which municipal and regional actors use TDPs to balance economic development and environmental or social pressures from tourism, there are nonetheless many good examples in the rural TDPs where sustainable tourism concerns are well integrated into larger societal development aspects.

Our analysis addresses the question of whether openness about negative as well as positive impacts of tourism in TDPs (either at the local level, or as something to learn from, from other destinations, and prevent in the local context) influences the way in which sustainability concerns are handled in the subnational TDPs. It shows that having experienced negative consequences of tourism, faced them and *expressed them in a TDP* is associated with having a more holistic view of sustainability. In our reading, including both negative and positive aspects of tourism in a TDP therefore indicates a willingness to learn from past mistakes and to plan for better future development, or what Kauppila *et al.* (2009) could call a 'tourism as a problem' discourse. We think that the results suggest that inclusive debates about future tourism development are important, and this requires an increased openness about both positive and negative effects of tourism (see also Kauppila *et al.*, 2009; Higgins-Desbiolles *et al.*, 2019). This is also useful for destinations that have not (yet) experienced negative effects of tourism, to plan for how to prevent or reduce these in a proactive manner, instead of having to come up with reactive responses.

We also asked whether the level of participation in the process leading to a TDP influences the resulting sustainability concern handling expressed in the TDP. Our analyses show that the more participatory and inclusive processes for the TDP development are associated with a more holistic treatment of sustainability concerns in TDPs. We find that there is a statistically significant association between inclusive, participatory TDP processes and those TDPs that treat sustainability concerns in a holistic, detailed and (relatively) operationalized way. As the process pre-dates the resulting TDP, we argue that this is likely to be a causal effect.

This important finding provides a basis for recommending that broader and more participatory processes must be adopted if future TDPs are to treat sustainability concerns more holistically. We also give further empirical support to calls for collective action among broad stakeholder groups, and for a reorientation of tourism towards the rights and interests of local communities (Fodness, 2017; Higgins-Desbiolles *et al.*, 2019), in line with the arguments by Bramwell and Lane (2010), Bramwell (2011) and Ruhanen (2013) for the importance of broad participation, which is, in the words of Kauppila *et al.* (2009), a prerequisite for successfully using 'tourism as a tool for regional development'.

The identified patterns suggest that those TDPs developed through broader participation and through collaboration with different sectors are positioned to better align broader societal interests. These TDPs do a better job of representing the region's agriculture and local food production, cultural life, physical environment and appearance, natural environment and biodiversity, history, authenticity, etc., and often result in a more holistic view on sustainability. Furthermore, the national differences in the way sustainability is addressed by subnational TDPs indicate that strong national goal-setting, followed up by institutional support (for innovation and benchmarking, among others), are important for helping the subnational TDPs set concrete goals for the tourism sector's contribution to sustainable development, including sustainable rural destination development.

Interrelating different national, regional, and local plans that influence, or are influenced by, tourism, and integrating them with rural development plans and in TDPs is challenging (Árnadóttir, 2019). Adding to this challenge, recent research recommends that adaptive management approaches should be implemented in destination management, and that they should include a wide range of stakeholders in planning processes (Øian *et al.*, 2018). Such approaches enhance the legitimacy of policy development, while including as many different aspects and opinions as possible. This should minimize constraints from the economic, social or political dimensions of sustainability. We concur with Luthe and Wyss (2014) that adaptive processes and network-based governance structures are also needed to address resilience in tourism planning processes.

Finally, our results suggest that introducing broad, participatory planning processes is essential for developing successful tourism plans. Additionally, openness about both the positive and negative impacts of tourism promotes a more holistic discussion about future tourism development. Many regions and municipalities did not take this approach when developing the TDP analysed in this study. Those that did so often ended up with a more sustainably inclusive and holistic end product. Therefore, including a diverse group of stakeholders in the planning process, and ensuring learning from both positive and negative experiences, should be encouraged to ensure broad inclusion of different aspects relevant to destination plans. There is still room for improvement in all of this, especially if tourism in the Nordic rural areas is going to transform itself into sustainable tourism and help rural Nordic tourism destinations develop in a more sustainable way.

### 9.6.1 Limitations

There are three main limitations of the study. The first is the breadth of the types of documents included in the study, in order to ensure a sufficiently wide geographical coverage. It is fair to assume that the specific type of 'author' of the strategic documents influences the context in which tourism (and its desired development) is seen and presented. For example, a municipal or regional authority planning document is more likely to include wider societal concerns than documents from sectorial-focused actors that have tourism development as their focal point. Consequently, the breadth of document types included in the TDP analysis may blur our results. Ideally, separate subanalyses should be carried out by type of author.

The second limitation is the retrospective nature of the study, because as soon as a strategy or a plan is made, it can in some ways be viewed as a historical document. Sustainability concerns are gaining attention within tourism, judging from media reports and research publications. We also notice that local actors engaged in developing new TDPs put sustainability concerns at a more central place than it was given in the expiring ones. It would be interesting to repeat the study of sustainability concerns in subnational tourism development plans in a few years.

Finally, the study is limited to a desk analysis. Therefore, it does not investigate the different ways in which strategies and plans were, or were not, implemented.

### 9.6.2 Directions for future research

Our conclusions indicate that more inclusive and participatory processes influence the degree to which concerns about sustainability and other issues go beyond the tourism sector. Consequently, we believe that it is important to further explore these planning processes. The Covid-19 pandemic, as well as concerns about $CO_2$ emissions from intercontinental air travel, have changed the tourism market, making 'staycation' and domestic tourism segments gain importance. How these changes can be used as an opportunity to plan for more sustainable tourism and to strengthen the synergies for the local population in the rural areas of the Nordic region are considered to be important areas of future research.

We see a potential – still nowhere fully tapped – for a more strategic use of tourism and hospitality-based plans, networks and initiatives integrating more sustainable tourism experiences and products into local and regional development. Future research into this, and the conditions that enable it, is also important from a planning point of view.

## References


Amore, A., Prayag, G. and Hall, C.M. (2018) Conceptualizing destination resilience from a multilevel perspective. *Tourism Review International* 22(3), 235–250. DOI: 10.3727/154427218X15369305779010.

Árnadóttir, R. (2019) *Nordic Tourism Policy Analysis*. Nordic Council of Ministers Secretariat, Copenhagen.

Bogason, Á., Karlsdóttir, A. and Broegaard, R.B. (2020) *Planning for sustainable tourism in the Nordic region. Pan-Nordic analysis of Regional Tourism Strategies for rural areas*. Nordregio, Stockholm.

Bramwell, B. (2011) Governance, the state and sustainable tourism: A political economy approach. *Journal of Sustainable Tourism* 19(4–5), 459–477. DOI: 10.1080/09669582.2011.576765.

Bramwell, B. and Lane, B. (2010) Sustainable tourism and the evolving roles of government planning. *Journal of Sustainable Tourism* 18(1), 1–5. DOI: 10.1080/09669580903338790.

Bramwell, B. and Lane, B. (2011) Critical research on the governance of tourism and sustainability. *Journal of Sustainable Tourism* 19(4–5), 411–421. DOI: 10.1080/09669582.2011.580586.

Broegaard, R.B. (2020) Rural destination development contributions by outdoor tourism actors: A Bornholm case study. *Tourism Geographies* 58(4), 1–21. DOI: 10.1080/14616688.2020.1795708.

Costa, J., Montenegro, M. and Gomes, J. (2018) What challenges and opportunities will lead to success? *Worldwide Hospitality and Tourism Themes: Tourism in 2030* 10(6), 631–634. DOI: 10.1108/WHATT-08-2018-0053.

Fodness, D. (2017) The problematic nature of sustainable tourism: some implications for planners and managers. *Current Issues in Tourism* 20(16), 1671–1683. DOI: 10.1080/13683500.2016.1209162.

Hall, C.M. (2009) Degrowing tourism: décroissance, sustainable consumption and steady-state tourism. *Journal of Sustainable Tourism* 20(1), 46–61.

Hall, C.M. (2011) A typology of governance and its implications for tourism policy analysis. *Journal of Sustainable Tourism* 19(4–5), 437–457. DOI: 10.1080/09669582.2011.570346.

Hall, C.M. (2019) Constructing sustainable tourism development: The 2030 agenda and the managerial ecology of sustainable tourism. *Journal of Sustainable Tourism* 27(7), 1044–1060. DOI: 10.1080/09669582.2018.1560456.

Higgins-Desbiolles, F., Carnicelli, S., Krolikowski, C., Wijesinghe, G. and Boluk, K. (2019) Degrowing tourism: rethinking tourism. *Journal of Sustainable Tourism* 27(12), 1926–1944. DOI: 10.1080/09669582.2019.1601732.

Innerhofer, E., Fontanari, M. and Pechlaner, H. (2018) Destination resilience. In: Innerhofer, E., Fontanari, M., and Pechlaner, H. (eds) *Destination Resilience: Challenges and Opportunities for Destination Management and Governance. Contemporary Geographies of Leisure, Tourism and Mobility*. Routledge, London & New York, pp. 1–208. DOI: 10.4324/9780203701904.

Kauppila, P., Saarinen, J. and Leinonen, R. (2009) Sustainable tourism planning and regional development in peripheries: A Nordic view. *Scandinavian Journal of Hospitality and Tourism* 9(4), 424–435. DOI: 10.1080/15022250903175274.

Luthe, T. and Wyss, R. (2014) Assessing and planning resilience in tourism. *Tourism Management* 44(3), 161–163. DOI: 10.1016/j.tourman.2014.03.011.

Müller, D.K. (2013) Rural tourism in the Nordic countries – a short assessment. In: Fredricsson, K. and Smas, L. (eds) *Small-scale Tourism in Rural Areas – Trends and Research in the Nordic Countries*. Nordic Working Group 1B: Future Rural Areas. Working Paper 3/2013, Nordregio, Stockholm, pp. 15–23.

Müller, D.K. and Jansson, B. (eds) (2007) *Tourism in Peripheries: Perspectives from the Far North and South*. CAB International, Wallingford, UK.

Øian, H., Fredman, P., Sandell, K., Sæþórsdóttir, A.D., Tyrväinen, L. *et al.* (2018) *Tourism, Nature and Sustainability. A review of policy instruments in the Nordic countries*. TemaNord 2018:534. Nordic Council of Ministers, Copenhagen.

Ruhanen, L. (2013) Local government: Facilitator or inhibitor of sustainable tourism development? *Journal of Sustainable Tourism* 21(1), 80–98. DOI: 10.1080/09669582.2012.680463.

Torres-Delgado, A. and Saarinen, J. (2014) Using indicators to assess sustainable tourism development: a review. *Tourism Geographies* 16(1), 31–47. DOI: 10.1080/14616688.2013.867530.

# Conclusion

**Dimitrios Stylidis**
*University of Crete, Greece*

This book has aimed to examine tourism planning and development in Western European countries, which constitute the most visited subregion globally. The edited volume included nice cases covering the following countries: Belgium, Denmark, the Faroe Islands, Finland, France, Germany, Iceland, Ireland, Italy, Malta, Norway, Portugal, Spain, Sweden and the UK, focusing on various issues, challenges and success stories of tourism development, planning and impacts in this subregion of the world.

Peter Wiltshier in the first chapter provided a critical evaluation of the purpose and value of sustainable development goals (SDGs), using the case of a small market town in the English Midlands. In collaboration with university students and public and private sector stakeholders – and using focus groups and visitor surveys, business stakeholders' meetings and visitor capacity activities – a marketing, communication and media plan for the destination was developed and tested. More precisely, via several evaluations, including a discourse analysis, an agreed brand and identity were proposed, drawn from the identified shared heritage. Within this informative rejuvenation and regeneration project, the value of new learning in bridging social capital is notable. The study contributed to tourism theory and practice by offering recommendations to small-scale destinations in mature places, that commonly lack skills, capacity and resources, highlighting the need for institutional innovation and a collaborative approach.

In the second chapter of this volume, Anna Serena Vergori investigated the mode of transport that international tourists used to arrive in Italy over the years 1997 to 2019, aiming to establish a relationship between mode of transport and the seasonal pattern of tourism demand. Following the deregulation of the air travel industry in the EU, the travel mode has radically transformed the past 20 years, with a notable increase in the use of air transport (mainly due to reduced cost and increased availability) over the other modes of transport, such as cars. Further, such changes in visitors' travel habits potentially lead to reductions in seasonality, highlighting new policymaking tools that aim to tackle this challenging issue that most destinations are facing worldwide. Policy interventions are proposed in the chapter by Vergori, including price incentives aiming to encourage air carriers to increase the number of flights during the off-peak season. Despite the prominent negative impact of the pandemic on the number of foreign visitors, the 'one-peak' seasonality pattern has persisted, indicating the habitual nature of tourist holidays in Europe.

The third chapter, authored by Marie-Louise Mangion, dealt with tourism development in the island of Malta, which is linked to high population density, urbanization and heavy dependence on international tourism activity. Via content analysis of the available development plans and tourism policy documents published

 *Tourism Planning and Development in Western Europe* (eds K. Andriotis *et al.*)
DOI: 10.1079/9781800620797.conc

in 1959 onwards, five tourism development themes were identified: growth, national prosperity, governance, the tourism offer and market orientation. A conceptual framework including various factors supporting or undermining short- and long-term tourism development was proposed. The analysis also revealed that there are pertaining issues of concern that are of a systemic character, first observed in the early days of tourism, which are still significant policy issues nowadays. Some omissions in these plans observed included lack of cooperation with the third sector, lack of residents' input in the tourism policy and planning process, and unequal distribution of income.

The next chapter, by Ana Domínguez-Quintero and Sonia Ancio Alcón focused on industrial tourism in the province of Seville, Spain, and explored the perceptions industrial firms hold of the benefits and costs of this type of tourism. Based on survey responses provided by 38 operating companies that accept tourist visits at their premises, it became evident that the vast majority of them belong to the agricultural sector. Most such operators receive tourists, individuals or groups, and families. Around seven out of ten organize tastings and offer their products for sale during the visit. It is notable that the volume of operators that accept visits in Seville has doubled in the last ten years, showcasing the increasing demand for such tourism products. The researchers concluded that firms offering industrial tourism activities get benefits, such as improved company image, acquisition of new customers, increased sales and a greater understanding of customer needs. Relative costs were considered low and most companies demonstrated a strong intention to maintain visits in the future. Despite the benefits of industrial tourism – a sustainable type of tourism with strong potential for both companies and the local community alike – policymakers commonly fail to pay attention to this activity.

Chapter 5, by Margarida Ferreira da Silva, Helena Albuquerque, Filomena Martins and Gildas Buron, explored salt pans as natural and cultural heritage attractions. The authors compared two cases, those of Aveiro, Portugal and Guérande, France, representing different models of development, the former being an area targeted on the exploration of salt pans for tourism, while the latter represents a well established salt pan with tourism serving as a complementary sustainable activity. The study aimed to draw conclusions and provide recommendations to the former following the best practices implemented in the latter. Based on data collected via document analysis of existing policy documents, the authors critique initiatives that are market-driven, exclusively focusing on tourism purposes that do not respond to the growing threat of traditional activity disappearance. Without strategies involving a set of stakeholders aiming to preserve and further develop the activity of artisanal salt-making in a sustainable context, these vulnerable cultural and heritage attractions are endangered.

Laura Rauschen, Anastasia Traskevich and Martin Fontanari, in Chapter 6, conceptualized the notion of micro-destinations and explored their relevance for tourism development and planning, using the case of East Belgium. The methodology was based on an interview with a representative of the local tourism administration and a workshop with the participation of members of local communities, aiming to obtain views on the implementation of sustainable tourism at the micro-destination level. The key defining characteristics of a micro-destination identified by the authors involved a unique and diverging culture, the homogeneity of locals, authentic history and nature, along with its particularity as compared to nearby areas. Following a bottom-up approach, the study highlighted the importance of local community participation and stakeholder engagement at the micro-destination level for the sustainable development of tourism.

Focusing on urban tourism and its impact on two mature European destinations (Milan, Italy and Porto, Portugal), Valeria Fusè and Elisabeth Kastenholz, in Chapter 7, identified similarities and differences in residents' perceptions of tourism impacts along with the nature and form of host–guest interactions in such contexts. Most of these interactions in both settings were conducted in public spaces and transport, and in facilities such as cafés, restaurants and shopping outlets. Following in-depth interviews, local residents evaluated such interactions with tourists as mainly positive and rewarding, enabling them to understand foreign cultures, and producing feelings of enthusiasm and stimulation. Such encounters appeared more often in the city of Porto than in Milan, and are related to the proportion of job dependence on tourism (higher in Porto). Residents in both cities showcased a consensus regarding the benefits of urban transformation via tourism development, which in the case of Porto is mainly limited to the old town, while in Milan it takes place in a more dispersed, spatial manner.

Valeria Diaferia, Anastasia Traskevich and Martin Fontanari, in Chapter 8, assessed and compared the sociocultural dimensions of destination resilience in Germany, Italy and Ireland. Using a blend of qualitative and quantitative indicators to assess these countries' resilience (emotions and creativity; learning skills and professional efficiency; inner and outer world balance; communication and positive thinking; ethics and values; spirituality), the authors developed suggestions for tourism product development in the countries under examination. Results revealed that Germany and Ireland have the potential to overcome crises, while Italy appears more vulnerable to shocks. The authors have contributed to the identification of elements of tourism competitiveness, attractiveness and sustainability, thereby paving the way for new methodological tools in tourism planning and development.

The last chapter, authored by Rikke Brandt Broegaard, Ágúst Bogason and Anna Karlsdóttir, focused on the diverse visions and goals for sustainable tourism development in a variety of Northern European countries including Denmark, Finland, Iceland, Norway, Sweden and the Faroe Islands. The authors conducted content analysis in 110 rural municipal, subregional and regional tourism plans, disclosing different approaches via which sustainability concerns are reflected in the different countries. Their findings indicated that most of the plans were short- (up to four years) or medium-term (five to eight years), developed at the regional level. Country differences were also observed, as Denmark had the highest number of plans developed at the municipal level, while the vast majority of the plans in Iceland and Norway were at the regional level. In all plans, economic growth (e.g. increase in the number of visitors) was the most sought-after benefit of tourism. The notion of sustainability was embedded in some plans, while there were cases in which it was not mentioned at all. About half of the plans in general adopted a holistic approach, considering the social, environmental and economic dimensions of sustainability. The authors concluded that more participatory processes and collaboration with different sectors must be adopted in planning and policymaking to treat sustainability concerns more holistically.

Overall, this book contributes to a better understanding of tourism planning and development in the Western European region, highlighting the key role public planning and policymaking play at the national, regional and local levels. The volume in particular emphasizes the value of direct community participation in such processes, at a local or micro-destination level. Notable also is the need to preserve and sustainably develop the cultural heritage and industrial activities that serve as unique selling points and are part of the identity and image of many European destinations. Several of the authors in this volume also highlighted a myopic approach towards the economic benefits of the industry, with much lower attention given to the sociocultural and environmental costs. Most of the key challenges identified here appear quite distinct from those highlighted (i) by Monterrubio *et al.* (2020) in Latin America, where the main issues were related to tourism dispersal, conflicts, unequal development and the underrepresentation of various ethnic groups in tourism decision making; and (ii) by Stylidis and Seetanah (2021), where the relevant tourism development challenges involved terrorism and crime, insufficient accessibility, weak institutional frameworks and overdependence on tourism along with fragile ecosystems. These results support the need for continuous research and understanding of the unique peculiarities and characteristics of tourism in the various subregions of the world.

While this book extends current knowledge on tourism development and planning in Western Europe, additional work in this region is required. Despite the editors' attempts to include all countries in the subregion, there was a lack of contributions from countries such as Austria, Greece and the Netherlands. Somewhat similarly, as each chapter focused on a particular tourism development issue related to the country under consideration, others might have been neglected. Next, most of the research included in this edited volume was conducted in the pre-Covid-19 era – although the authors have attempted to link their results to the current scenario, additional data are required to better reflect this period and the unknown future of tourism development worldwide.

## References

Monterrubio, C., Andriotis, K. and Stylidis, D. (2020) *Tourism Planning and Development in Latin America*. CAB International, Wallingford, UK.

Stylidis, D. and Seetanah, B. (2021) *Tourism Planning and Development in South Asia*. CAB International, Wallingford, UK.

# Index

Note:- Page numbers in **bold** type refer to figures and page numbers in *italic* type refer to tables.